Telecentres, Access and Development

Experience and Lessons from Uganda and South Africa

Telecentres, Access and Development

Experience and Lessons from Uganda and South Africa

Sarah Parkinson

ITDG
PUBLISHING

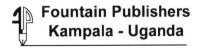

Fountain Publishers
Kampala - Uganda

International Development Research Centre
Ottawa • Cairo • Dakar • Montevideo • Nairobi • New Delhi • Singapore

Published by ITDG Publishing
Schumacher Centre for Technology and Development
Bourton Hall, Bourton-on-Dunsmore, Warwickshire CV23 9QZ, UK
www.itdgpublishing.org.uk
ISBN 1 85339 619 2

and Fountain Publishers
P.O. Box 488, Kampala, Uganda
www.fountainpublishers.co.ug/fountain@starcom.co.ug
ISBN 9970 02 517 1 (Africa only)

and the International Development Research Centre
P.O. Box 8500, Ottawa, ON, Canada K1G 3H9
www.idrc.ca/pub@idrc.ca
ISBN 1 55250 189 2 (e-book)

First published in 2005

A catalogue record for this book is available from the British Library.

ITDG Publishing is the publishing arm of the Intermediate Technology
Development Group Ltd. Our mission is to build skills and capacity of people
in developing countries through the dissemination of information in all forms,
enabling them to improve the quality of their lives and that of future generations.

Cover photographs courtesy of the International Development Research Centre.

Typeset by RefineCatch Limited, Bungay, Suffolk
Printed in India by Replika Press

Contents

Foreword

The telecentre movement is more than two decades old. What started as an attempt to help farmers in rural Sweden[1] learn more about their financial accounts from spreadsheets has morphed into a worldwide movement. And 'movement' is the right word rather than institution, organization or system.

The source fuel for the people that start, manage and grow telecentres is their service to the community and the skills they learn and grow. Telecentres have mushroomed into worldwide phenomena because of the unbridled advocacy for their usefulness that those who have had the telecentre experience possess. This experiential learning and professional passion is difficult to share without the actual experience of helping a telecentre and the movement succeed.

What is the role of research and the researcher in this type of movement? Does the researcher stand outside the movement and offer detached empiricist observation disconnected from the actual development experience? This volume illustrates the intellectual movement of its author from that of abstracted positivist to applied telecentre partisan. Her research isn't aimed at a disinterested academy where the goal is only professional promotion. It is focused as much on helping telecentre practitioners understand from one another's experiences and, equally, building a bridge to those who don't yet understand, and should.

Because the telecentre movement is without the normative patterns of more traditional institutions, it is sometimes difficult for academics to understand what they are observing. As they sit outside the tacit knowledge, informal learning, unstratified relations and networked knowledge dissemination, the culture of the telecentre and its movement can seem disorganized and in need of 'professional' guidance. It is only by entering the gestalt of the digital development experience the telecentre represents that understanding of its assets and its needs for further development can really be achieved.

As a young professional researcher, the author has managed this in a

---

[1] The first rural telecentre, or telecottage as it was then called, was established in Velmdalen, Sweden in 1985 by Henning Albrechtson.

remarkable way. Her volume speaks to the commitment of the telecentre activists she has worked with and observed. It also speaks to the fact that this she, and this volume, have now become a part of the telecentre movement.

Rich Fuchs
Director
Information and Communications Technologies for Development
International Development Research Centre

Acknowledgements

As both an institution and a collection of talented individuals, IDRC provided me with a valuable professional opportunity, support and guidance. For this I am profoundly grateful. In particular I wish to thank Richard Fuchs, Laurent Elder, Morenike Ladikpo, Steve Song, Heloise Emdon, Bill Carman and the IDRC Awards Office.

My research assistants were also crucial to this work. They provided field support in many ways, especially by carrying out the bulk of household surveys, serving as interpreters, and offering insight and advice. They are: Millicent Poni and Monica Mraji in Khayalitsha; Simon Mkhize and Sbusiso Gcaba in Bhamshela; Paul Ojara, Agnes Ekwar, Mary Omoko, Silvia Otto and Grace Aguma in Lira; Caroline Korishaba Nkunda, Patience Tumukunde and Boaz Agaba in Kabale; and William Ssemakula and Geoffrey Kikomeko in Nabweru.

Many others in Uganda and South Africa also supported this research by giving advice and valuable information during interviews. Although too numerous to mention by name, they are warmly remembered.

A final note of appreciation goes to Gerry Toomey of Green Ink Publishing Services Ltd. for his great editing.

Sarah Parkinson
April 2005

Tables

Figures

Photos and boxes

Acronyms and other abbreviations

ADP	Area Development Programme
ADSL	asymmetric data subscriber line
AHI	African Highlands Initiative
ANC	African National Congress
AT&T	American Telephone and Telegraph Company, now referring to itself simply as AT&T
CBO	community-based organization
CD-ROM	compact disk, read-only memory
CEEWA	Council for Economic Empowerment for Women in Africa
CLC	Community Learning Centre
CMS	Community Multimedia Services
COSATU	Congress of South African Trade Unions
CPAR	Canadian Physicians for Aid and Relief
CSO	civil society organization
CSOs	community service obligations
CV	curriculum vitae
DACST	Department of Arts, Culture, Science and Technology
DFID	Department for International Development, UK
DoC	Department of Communications
DoE	Department of Education
GATS	General Agreement on Trade in Services
GCIS	Government Communication and Information System
GDP	gross domestic product
GSM	Global System for Mobile Communications
HQ	headquarters
HS	high school
ICASA	Independent Communications Authority of South Africa
ICT	information and communication technologies
IDRC	International Development Research Centre
IMF	International Monetary Fund
ISETT SETA	Information Systems, Electronics and Telecommunications Technologies Sector Education and Training Authority

ISP	Internet service provider
ITU	International Telecommunication Union
KDF	Khayelitsha Development Forum
KERIC	Khayelitsha Educational Resource Information Centre
LAN	local area network
LRA	Lord's Resistance Army
MDDA	Media Development and Diversity Agency
MPCC	Multipurpose Community Centre
MTN	Mobile Telephony Networks Ltd.
MTV	Music Television
NAADS	National Agricultural Advisory Service, Uganda
NCRF	National Community Radio Forum
NGO	non-governmental organization
NITF	National Information Technology Forum
OWN	Open Windows Network
PEAP	Poverty Eradication Action Plan
PIT	public information terminal
PNC ISAD	Presidential National Committee on the Information Society and Development
PoP	Point of Presence
PPA	Participatory Poverty Assessment
RANET	Radio and Internet (international programme)
RCDF	Rural Communications Development Fund
RDP	Reconstruction and Development Plan
RSA	Republic of South Africa
SACRIN	South African Information Network
SATRA	South African Telecommunications Regulatory Authority
SBT	school-based telecentre
SMME	small, medium and micro enterprises
SMS	Short Message Service
SPSS	Statistical Package for the Social Sciences
SS	secondary school
TASA	Telecentre Association of South Africa
telecom	telecommunication(s)
UCC	Uganda Communications Commission
UNCST	Uganda National Council for Science and Technology
UNESCO	United Nations Educational, Scientific and Cultural Organization
USA	Universal Service Agency
USB	Universal Serial Bus
USAID	United States Agency for International Development
USF	Universal Service Fund
UTL	Uganda Telecommunications Ltd.
VoIP	voice over Internet protocol
VoK	Voice of Kigezi (Radio Station, Kabale)

VSAT	very small aperture terminal
WOUGNET	Women of Uganda Network
WTO	World Trade Organization

Executive Summary

Introduction

This study examines the role that shared access centres play in government strategies to provide universal access to information and communication technologies (ICTs). It also aims to shed light on the external factors that affect the performance of these centres. The analysis draws on the experiences of South Africa and Uganda in 2003, at the policy and community levels. Both countries have policies regarding universal access and have taken steps to achieve their policy goals.

In this study, shared access centres are understood to exist in the context of a broader universal access policy, which is itself embedded in a national development policy framework. They are affected, and sometimes directly created, by policy and implementation processes, and involve a variety of actors at the local, national and international levels. Key actors include licensed telecommunication operators, regulators, donor agencies and local professionals.

This research examined the situation in selected urban and rural communities in Uganda and South Africa. The more than 50 centres that were reviewed in the course of the study are classified into seven types:

- phone shops;
- Internet cafés;
- communication centres (including computer training centres);
- services within an existing business (eg, a hair salon, traditional healer or stationery shop);
- services within an existing not-for-profit organization;
- telecentres;
- centres that are part of an integrated development strategy.

Four factors are posited as important to the success of shared access centres in achieving universal access and maximizing development potential: scalability, sustainability, reach and use. They cumulatively contribute to the development impact of such centres. The apparent impact of access centres is analysed in two ways. Firstly, the situation in communities is compared with the goals motivating universal access policy in each country. Secondly, characteristic usage of different access centres is analysed in the context of the livelihood strategies commonly used by different people within the case communities.

Key findings and recommendations

The key findings from this report are presented below as 11 lessons, each presented briefly with some evidence and explanation, and specific recommendations where appropriate.

Lesson 1. Access centres are generally not adequate by themselves to build local demand.

Evidence/rationale
Usage patterns at telecentres in both South Africa and Uganda show that certain services, especially those related to computers and information services, tend to be poorly utilized, whilst some others, including telephone and photocopying, may enjoy moderate to high levels of demand. Similarly, private enterprises have greater success when offering the latter services rather than the former. This is especially true in more rural communities, and demand within these underused services does not show any tendency to increase over time. Rather, these services tend to be difficult to maintain over time, especially where telecommunication and electricity services are expensive and/or erratic. People using computer-related services tend to have higher levels of education (secondary or above) and often have had some prior exposure to computers through school, work, or are introduced through a friend or family member. Unless people have other venues for building their awareness of and confidence in using ICTs, telecentres have not proven a robust method of overcoming the multiple barriers to access that many people face.

Recommendations
Policy makers should focus on creating an enabling environment for access centres, especially by focusing on liberalization of the national telecommunication market. They should also recognize that, with the use of computers and the Internet, people need time and space in which to learn basic skills, and that fee-based access is not a viable way to achieve this. Computers in schools may be one option, although this solution has its own set of challenges. Those with the task of setting up centres need to first assess the local situation, not just needs but also feasibility and potential risks.

Lesson 2. Because access to ICTs is difficult to achieve in rural areas through market means, and may not by itself be a primary development goal, it should, where relevant, be carefully harmonized with larger rural development strategies.

Evidence/rationale

Stand-alone access centres are difficult to maintain in rural areas and tend to have limited market demand, with the exception of phone services. This also means that, even when externally supported, they tend to have limited use and impact. Integrating ICTs into larger rural development projects means that these technologies can contribute to overall development goals while the projects themselves make ICT service provision more feasible, although still dependent on overall national market structures. This approach can support ICT services by providing an anchor market, stimulating demand and helping people to locate and apply useful information. South Africa and Uganda have large national programmes that may demonstrate this lesson, but they are still in too early a phase to permit comment on their practical success. RANET (Radio and Internet) Uganda, a programme to deliver climatic information to farmers, has demonstrated success through building on the existing infrastructure of World Vision's area development offices.

Recommendations

Stand-alone access centres should be pursued as an option only where local markets have demonstrated capacity to support them. Otherwise, ICT access should be integrated into other development activities. How this is done will depend very much on national development priorities and programmes. This approach should be viewed as social investment; it may take time and will require careful planning. It may also benefit from cross-sectoral collaboration within government and partnerships between government and external stakeholders, including NGOs and private sector service providers.

Lesson 3. Any attempt to implement ICT access centres will be strongly constrained (or enabled) by the national telecommunication market.

Evidence/rationale

For example, in South Africa the attempts of the Universal Service Agency to institute self-sustaining community telecentres have been severely undermined by the high cost and poor service provided by the incumbent telecommunication operator. Had greater technological choice and market competition been available, some common problems would have been avoided and the telecentres would have had a much greater chance of long-term survival. Thus, it was a strategic error, at the national level, to put resources and effort into implementing access centres when they depended on a single provider whose services they could not afford.

Recommendations

This lesson reinforces the need for national telecommunication market liberalization and regulation. It also suggests that government should not consider direct implementation of access centres without addressing these issues. Likewise, third parties may be more effective as consumer advocates and lobbyists where vested interests are preventing reforms that would be in the public interest, rather than creating expensive stand-alone centres that are neither sustainable nor replicable.

Lesson 4. Universal access for telephone requires strategies different from those needed for other ICTs (especially computer-related), since the former usually enjoys immediate demand while the latter does not.

Evidence/rationale
In both South Africa and Uganda, telephone services were generally well used and appreciated where they were available, whereas other ICT services were generally underutilized and did not have ready local demand. This meant that, in many cases, stand-alone telephone access could be sustained through user fees, whereas other types of ICT services were much more difficult to sustain. Awareness of computers and the Internet was generally lower than that of telephone service, thus presenting obvious extra barriers to the use of these technologies. Another explanation is that communication is widely valued and appreciated more than information-related services, an observation further reinforced by the fact that the Internet, where accessed, is used primarily for e-mail.

Recommendations
Telephone access and use are issues best considered separately from Internet access and use, especially since Global System for Mobile Communications (GSM) cellular networks have contributed broadly to the spread of the former. Nevertheless, policy makers should maintain a holistic view that considers the possible relationship between the two: trends of technological convergence, especially voice over Internet protocol (VoIP), may again reduce the distance between these two in terms of access, if not usage, if telephone demand becomes a driver for Internet infrastructure.

Lesson 5. ICT access centres are an important but politically weak part of the access chain. They often get short-shrifted in terms of service quality and cost. This undermines their chances of long-term survival and acts as a disincentive to the creation of such centres in the first place.

Evidence/rationale
Sustainability is often a struggle for both public and private access centres in both countries, especially in South Africa. Public access providers must buy their services from a limited number of licensed telecommunication operators. Prices are already high and they cannot mark them up for consumers. This means their profit margins are often dangerously low even where local demand is present. A fairer approach is Vodacom's community service programme, whereby it gives preferential rates to entrepreneurs running its phone shop franchises. These entrepreneurs make good profits while offering public services at local market rates, adding significantly to the sustainability, rapid

spread and community development value of these enterprises. This is a model that should be more widely encouraged by public policy.

Recommendations

Regulators should consider encouraging or supporting preferential tele-communication service rates (and possibly electricity rates) to agencies, whether public or private, which provide public access.

Lesson 6. Affordability is still a major barrier to accessibility in both countries.

Evidence/rationale

In both South Africa and Uganda, unaffordable prices are cited by users, non-users and access providers as a major reason for the limited use of ICTs. The affordability problem is in turn attributed largely to lack of competition in the national market. For example, where cellular markets have proven much more lucrative than expected, there has not been a large drop in the consumer price of services (Tusubira et al., 2004). In South Africa, the high cost of landline telephones has increased at a rate that, according to analysts, exceeded any reasonable explanation other than the self-interested profit motive of the major operator and the government's failure to regulate in the public interest (Emdon, 2003). This in turn has led to very low rates of Internet use, even in Khayelitsha, one of the largest townships in the country.

Recommendations

Since much of the affordability problem relates to lack of competition, the highest priority is to liberalize the market, and especially to allow different technologies, such as VoIP. Independent regulation is obviously crucial. This may be aided in turn by the presence of organizations that can represent and advocate for the interests of consumer and small business organizations. The regulator, too, must be able to act in the public interest and consider where this prevails above the usual logic of the market – for example, through provision of free emergency calls to police or medical authorities.

Lesson 7. The popular view of ICTs casts computers as a tool for the educated only. This limits spontaneous appropriation of ICTs even where physical access is provided.

Evidence/rationale

This perceived link between computers and education was recorded in both South Africa and Uganda and came across strongly in all community household surveys, especially when people were explaining why they did not use computers. If one does not have an education and is not intending to get a formal job, there is little motivation to use computers. This conflicts with some policy assumptions that computers, and especially the Internet accessed through computers, can be tools to support broader democracy and bring the government closer to the people.

Recommendations

Avoid assumptions about how people will choose to use ICTs; any policy that depends on people's interaction with ICTs should be based on evidence. Otherwise, any development goal that counts on people using ICTs a certain way may also need to include the requisite social structures to encourage such use. South Africa's Multipurpose Community Centres (MPCCs), for example, are, according to the government agency in charge of co-ordinating them, explicitly intended to help people find information based on and relevant to their most general queries and concerns. Public information terminals, earlier implemented by the South African Department of Communications, were intended to help people obtain government information, but were not used much.

Lesson 8. Where ICTs are available and used, not all uses yield financial returns for their users or their communities. Thus, certain ICT uses, whilst benefiting large, often foreign telecommunication and equipment providers, may impede local economic development. For this reason, the presence of ICTs within communities cannot be presumed to indicate development, except where there is also evidence of some gain to the locality in which they are present.

Evidence/rationale

In both Uganda and South Africa, there is plenty of anecdotal evidence that cellular phone ownership is a major social status symbol. People, especially the young, are reported to be willing to pay a large proportion of their income and savings towards purchasing and maintaining a phone. (To a lesser extent, e-mail and Internet use is also a status symbol.) In Internet cafés, services are used primarily for e-mail, and secondarily for games and other entertainment which have no clear economic return. Again, this contradicts the optimistic national policy assumption that ICTs will be used in ways that stimulate development.

Recommendations

Since national universal access strategies are motivated by a desire for development and equal opportunity, they should be monitored through indicators which, going beyond mere access, are tied to national development priorities. Affordability of ICT services also needs to be defined by the national regulator according to national conditions.

Lesson 9. The relationship between the market and initiatives set up by external funders varies according to factors such as infrastructure and local economic activity. Market factors are often not well assessed or understood before a project is implemented. This can result in funded projects which, driven by their own sustainability requirements, emulate rather than complement their market counterparts.

Evidence/rationale

In Uganda, donor-funded telecentres were often slower to react to new market and technological opportunities than their private-sector counterparts. The African Highlands Initiative (AHI) telecentres in both Nabweru and Kabale had to disconnect their dial-up Internet connections when private competitors got cheaper broadband access. Neither managed to extend its user base significantly outside the same segment of the population that private centres catered to, and neither could meet its operating costs as had been intended by the project planners.

Recommendations

Any agency involved in implementing a not-for-profit access centre needs to assess the current and likely future market situation in the locale where the centre is to be located. Specifically, if the market is not active in the area, why not? It may be because of lack of demand, or because the infrastructure makes offering services far too expensive, or because entrepreneurs do not have start-up capital. Once the local context is understood, the implementing agents should be clear as to what their centre will accomplish that the market cannot, so that it does not become victim to the same limitations (eg, poor infrastructure) or simply end up offering the same services already available through the market. Partnership with and support for local entrepreneurs should be considered.

Lesson 10. Sustainability implies managing costs and complexity. Embedding ICT services within, and building upon, existing institutions is one of the best ways to do this. That strategy can also help maximize the development impact of services.

Evidence/rationale

World Links school-based telecentres in Uganda, the New Era Driving School in Khayelitsha, and RANET Uganda's partnership with World Vision are all examples of how the provision of ICT access has been sustained in large part by embedding it in an existing institution.

Recommendations

Those who implement access centres should consider collaborating with and, where possible, working within existing organizations in communities. Policy makers should consider ways to encourage such organizations to provide public access to their own ICT facilities where they exist. This can be done, for example, by easing licensing requirements and giving preferential rates or tax incentives. However, these organizations may have characteristics that will prevent them reaching the whole public, and these should also be taken into consideration. For example, schools are often slightly isolated and may be unwelcoming to illiterate community members.

Lesson 11. One of the biggest impacts of ICT use appears to be the maintenance of links between geographically dispersed family members. This may help to equalize other inequalities implied by the introduction of ICTs since family members with better education, ICT skills and related job opportunities usually share economic benefits with extended family members who would otherwise be unable to benefit.

Evidence/rationale
Telephones are most commonly used for communicating with family members. Significantly, this is a use that cuts across all sectors of society in both countries. For example, rural, largely subsistence-based family members use the telephone to communicate with formally employed relatives based in towns and cities. This communication facilitates the ongoing function of extended family support networks within and across countries. In Uganda, many families have overseas relatives and e-mail is likewise used by some of these as a cheap means of maintaining contact.

Recommendations
The role of within-country migration and international Diaspora communities in maintaining and contributing to rural development, and the role of ICTs in supporting it, is an area requiring greater attention from both policy makers and researchers.

CHAPTER ONE
Introduction

This book focuses on developing understanding around two key issues:

- the role that different models of shared access ICT (information and communication technology) centres have played in meeting universal access goals and supporting development in South Africa and Uganda;
- the external factors that have facilitated and constrained the activities of these centres.

Shared access ICT centres include cybercafés, telecentres, public phone shops and booths, computer secretarial services, and any centre, for profit or not, that provides shared public access to one or more ICTs.[1]

South Africa and Uganda were selected for the study because their political leaders had expressed commitment to the idea of using ICTs for development, and because both countries have had relatively early experience in putting such ideas into action. Each country chose a different path to implement its ideas, with much of Uganda's early experiences sparked by foreign donor interest. Thus, the depth and range of experience within these two countries provide a valuable opportunity for broader learning around these issues. The analysis is based on a review of national policy and implementation strategies, case studies of access centres in five communities (two in South Africa, three in Uganda), and analysis of the links between national policy and strategies and on-the-ground experiences. This analysis is intended to clarify the role that access centres play in larger universal access and development strategies. This is further used to identify and recommend strategies that policy makers, donor agencies and supportive intermediaries (such as national NGOs, networks and associations) can use to support and strengthen shared ICT centres and to increase their developmental impact.

The study had four research objectives:

- describing what universal access means and is intended to achieve, as stated by policy and national leaders in South Africa and Uganda;

- exploring what forms shared access centres have taken and their role in national universal access strategies;
- analysing the roles that different forms of shared access centres have played in development;
- identifying the ways policy makers, donors and supportive intermediaries can expand, sustain and use shared access initiatives to achieve universal access goals and maximize developmental benefits.

Key definitions

Shared access centres

Shared access centres, sometimes referred to as access points, are resources available to either all or a significant segment of the public. They provide direct access to ICTs and related value-added services. Common examples are public telephones, cybercafés, telecentres, computer training centres, and computer secretarial services and business centres.

Telecentres

'Telecentre' is a widely recognized term that encompasses most variants of shared access facility with an explicit development objective. Most telecentres are multipurpose, offering a range of services and ICTs, often including photocopying, computer typesetting, faxing, Internet (although many are beset by connectivity problems), phone and computer training, plus other value-added services that vary from site to site. Most telecentres in Africa have been established by external funding agencies in partnership with local NGOs and/or government structures. They are intended to 'bridge the digital divide' by reaching those who otherwise would be unlikely to access services. As such, telecentres are sometimes located in places where the market is not providing such services, or otherwise attempt to reach a broader segment of the population than the market would.

Information and communication technologies (ICTs)

The term ICTs usually refers to telecommunication-related and digital computer technologies, although with technological convergence, the distinction between telecommunication and broadcasting technologies is decreasing in both clarity and relevance. This research focuses specifically on telephone, fax, computer and Internet/e-mail, although it also includes a project that uses data and voice transmission over satellite broadcasts. While radio is also considered within the scope of the research, it is not considered an ICT but rather a 'close relative', while television and other media are not included. This follows general convention and the definitions used in most national access policies.

Universal access

Universal access is a common policy goal in which 100 per cent of the population is able to make use of a publicly available resource, whether it be basic schooling, health centres, or in this instance, ICTs. Universal access to ICTs is defined by some benchmark that changes from country to country and over time. For example, South Africa defined universal access as the population within 30 minutes of a public telephone, while Uganda defined it as a public telephone at every subcounty or community with a population of 5,000 or more. Once targets are reached, the benchmarks can be raised: for example, from within 30 to within 15 minutes of a telephone. Rapid technological change also requires that these definitions be regularly reviewed. Most policies attempt to be 'technologically neutral,' focusing on the service (ie, point-to-point voice telephony) rather than the means of achieving it. In practice, however, monopoly rights granted through licensing often limit the application of new technologies that could otherwise help to reach access targets.

Universal service

Universal service is a common policy goal whereby 100 per cent of the population is able to receive, by reasonable request and at a reasonable cost, a specific service on an individual or household basis. Traditionally, in the realm of ICTs, the target has been universal telephone service; that is, all households should be able to have, on request, home telephone service at an affordable rate. This definition has expanded in many countries to include value-added phone services such as messaging and voice mail as well as data transmission and Internet capacity. In countries with low average incomes and very low teledensities, universal service is not normally an immediate short-term policy goal, but may remain as a long-term goal.

Development

In this research, 'development' is understood in two senses. Firstly, it is progress towards whatever a national government has defined within its policy as the priority goals of its people.[2] Secondly, it refers to the increased capacity to choose. When applied to a community or national collective, this definition further requires that the capacity of one individual to choose does not reduce the range of freedom of others. Thus, it is an increase in the overall range of choice and freedom at the individual and collective level. This latter definition is based primarily on the work of Amartya Sen.

Context and rationale

Brief background on universal access in South Africa and Uganda

In recent years, universal access to ICTs has become a policy goal for many national governments, international development agencies and intergovernmental agencies such as those of the United Nations. ICTs are widely seen as a fundamental element of a newly emerging global information (or knowledge) society. As such, they represent both threat and promise. They may lead to greater opportunities for those who can partake of them; but they may also lead to greater exclusion for those who cannot.

The fear of this rift is made more urgent by the observation that new and deepening forms of exclusion are appearing in our globalizing society. In Africa, the development of nations has often been slow or stagnant and even when national indicators are moving up, close observers are often skeptical as to how much of this growth is reaching the poorest people and communities (eg, Jamal, 1998). In the context of rural Africa, the focus is not only on including people in the information society, but also on what these ICTs can offer individuals and communities who are in immediate need of better services and livelihood opportunities.

South Africa and Uganda are two African nations which made early and strong political commitment to the concept of universal access and whose leaders publicly requested the international community to help them in their efforts. By the mid-1990s, the governments of both countries were engaged in similar processes of developing universal access policies and setting up bodies to regulate telecom operators and implement access initiatives, while beginning to privatize and liberalize their telecommunication sectors. They were influenced and supported by a large number of international players and to a lesser extent by their own private and civil sectors. These two nations had unique experiences with universal access as a whole and with the role of shared access centres in particular. Their stories illustrate the relationship between shared access centres and broader universal access and telecommunication reform strategies, and the kinds of influence that national governments can exert, positively or negatively, on grassroots ICT access efforts.

The rhetoric of universal access is often underlain with self-interested political and economic forces redefining the public interest to coincide with their own. In particular, international telecommunication companies and the old state telecommunication monopolies are infamous for their talents as influential lobbyists on their own behalf. The American Telephone and Telegraph Company (now known simply as AT&T), coined the term 'universal service' in an argument to be granted monopoly rights in the USA in the early twentieth century (Melody, 1998). More recently, large telecommunication operators attempt to influence the international playing field and enter into new markets on terms that will maximize their own profits. 'Universal service' is still an argument that they use to justify monopolistic conditions that they claim are necessary to allow and justify large investment in telecommunication infrastructure.

The WTO in 1998 established rules of telecommunication sector reform in the General Agreement on Trade in Services (GATS). South Africa and Uganda were both signatories to this agreement. It requires signatory nations to set up an independent regulator while allowing them to determine their own definitions and indicators for universal service and access.

National policy makers and regulators play the crucial role of protecting the public interest through policy creation, implementation, monitoring and revision, and through the effective enforcement of regulatory conditions. On the one hand, failure to do so implies that universal access policy is simply rhetoric to sugarcoat the self-interested actions of powerful telecom operators, which are unlikely to coincide with public interest or to achieve universal access (Melody, 1998). On the other hand, success in creating publicly meaningful universal access policies that are integrated into a larger development strategy implies greatly increased opportunities for socio-economic development and improved well-being.

Shared access centres, universal access and development

Telecommunication reform has been undertaken in many African countries, including South Africa and Uganda. These processes follow the same general structure and logic, although heavy lobbying by various interested parties means that they do not always run quickly or smoothly. Interesting dramas are often created along the way.

Both South Africa and Uganda, like most African countries, previously had state-owned telecom monopolies with lacklustre performance, resulting in low teledensities, a large unmet demand for telephone service, low quality of service and little telecom infrastructure outside major urban centres. In addition, South Africa, although it had one of the highest teledensities in Africa at the end of the apartheid era, suffered a huge disparity in service between different racial groups.

The process of reform has generally involved partial privatization of the telecom company. Typically, one or more companies, mainly foreign-owned, buy shares while the government retains a significant number of shares which it is supposed to dispose of over time. The company is granted a period of exclusivity, usually about five years, coupled with licensing obligations related to performance measures and universal service targets. At the end of exclusivity, the market is opened to further competition, which may include one or more licensed operators. During the same period, cellular phone companies have entered the market and been licensed under a separate category. South Africa and Uganda each had three licensed cellular operators in 2003.

Both telecom sector reform and licensing obligations are intended to directly support universal service and access goals. By themselves, however, they are not sufficient to realize the full potential of ICTs for development. Bill

Melody argues that while these changes may increase the supply of telecom infrastructure, they are unlikely, given historical precedent, to fully penetrate rural areas. Even if they do, they will not bridge the accessibility divide unless demand can be built through greater public awareness of ICTs and capacity to use them (Melody, 1998).

This is where the potential role of shared access centres becomes apparent. Telecentres, for example, originated in Scandinavia as a way to make ICTs more economical, but more importantly, to build demand by sensitizing and training previously unexposed segments of the population to the potential utility of the technologies. The collective aspect of the centres also meant they could serve as focal points for community development. In the mid-1990s, many African countries were exposed to the telecentre concept through international symposia, where the International Telecommunication Union (ITU) championed them as valuable development tools. A number of international donor agencies set up pilot telecentre projects in Africa. And the South African government implemented an ambitious telecentre programme as part of its own universal access strategy. At the same time, entrepreneurs in African cities and towns began setting up cybercafés, telephone bureaux and business centres offering ICT services. While these were motivated by perceived market opportunities, not development goals, they often offered similar services, and sometimes competed directly with the telecentres.

Where this research fits in

This research aims to analyse and learn from on-the-ground experience with different forms of shared access centre in the context of overall universal access policy goals and strategies. These centres are generally recognized to be an important component in universal access, and especially in harnessing ICTs for development. However, it is not yet well understood how these centres contribute to, and are affected by, broader universal access strategies in the African context. Because most initiatives have been on a pilot scale, they have not yet been fully employed towards achieving universal access on a national scale. South Africa did make an ambitious attempt to do this, but unfortunately fell well short of the mark, as described later. However, these efforts, together with various pilot projects, stand-alone initiatives and market-driven entrepreneurial centres, provide an important chance to learn what role shared access centres could play, on a larger scale, in promoting universal access in a way that furthers development.

Methodology

The overall goal of the research was to understand, across multiple contexts, phenomena that are inherently dynamic and complex. Policies, markets, communities and projects are continually changing. So it was important to capture trends over time rather than merely take a snapshot of individual situations. Because there was no clear conceptual basis at the outset for deciding which elements should be the focus of attention, the research had to be somewhat open-ended.

The general methodology was to conduct five local case studies nested in two national case studies. Uganda and South Africa, as the national cases, allowed for comparison in terms of policies, market environments, and the apparent effects of these. The selected local communities in each country were analysed both as sub-components of the national studies and as case studies in their own right. These communities were selected based on two main criteria: firstly, they had an access centre that had been previously documented, and secondly, together with the other communities, they were representative of the spectrum of contexts (traditionally underserviced areas, both rural and disadvantaged urban) typically targeted by access initiatives. Within each community, the known access centre was researched with a view to under-standing change. In addition, other access centres and ICT services available within the area were identified and researched. These typically included about ten other centres, including cybercafés, computer-based secretarial and printing services, public phone shops, local radio stations and training centres. This research included a semi-structured interview with management or staff and was sometimes augmented by user exit surveys. A small survey of between 65 and 95 households was conducted in each area. In total, 62 access centres were included in the study and 370 households were surveyed.

The national case studies were augmented by primary and secondary research on national policy, regulation, ICT-related service providers, national government programmes and NGOs. Research in each country also included examples of local projects extraneous to the community case studies.

As with any research, the approach was also delimited by available time and resources. Fieldwork was carried out over three and a half months, with locally hired research assistants doing the bulk of the survey work, as well as serving as guides and cultural and linguistic interpreters. Research assistants also contributed to the design of the case research.

Field methods used

Data collection took place over a period of six and a half weeks in South Africa (April to June, 2003) and seven and a half weeks in Uganda (June to July, 2003). Field methods are described in Table 1.1. Appendix 1 is a list of what was included in data collection and from whom.

Data handling and analysis

Interviews were recorded through note-taking. Observations were noted and typed after the event. Questionnaires and surveys were recorded directly onto pre-made sheets. Any translation was done on the spot by the research assistants so that all information was recorded in English. Survey and

Table 1.1 Field methods

Method	Target group(s)	Description
Meetings	Key informants: other researchers, individuals with past involvement in various projects, etc.	Unstructured conversations
Semi-structured interviews	Access centre staff and management (also included community radio, high schools with computer labs) NGO representatives Government representatives Business representatives (esp. service providers)	These, along with secondary data review, were the primary data collection technique used at the national level and were also used extensively with key people at the local level
Access centre surveys	Staff and/or management at access centres	These were an abbreviated and more structured form of the access centre interviews
User exit surveys	Clients of access centres	These were standardized surveys across all cases and were conducted at access centres over the course of a day or a half-day, coupled with observation. Surveys were administered orally in the local language. Sample size varied from 2 to 20, depending on the volume of users
Student questionnaires	Those who were currently in a computer training course	Variant of user exit surveys
Household surveys	Households were randomly visited within 3 or 4 purposively selected neighbourhoods or villages, representing relatively rich, poor and average wealth settlements	Sample size from 65 to 95, structured questionnaires took about 20 minutes to complete and collected household and individual data on livelihood activities and ICT awareness and use
Observation	General operation and use of access centres	Notes and discussion amongst research assistants of observations made during other data collection
Document review	National level policy and analysis, previous history of one access centre in each community case study, other documentation as available/ relevant	Collection and review of existing documentation

questionnaire data were entered into an Access database and from there could be imported into Excel and SPSS (Statistical Package for the Social Sciences) for quantitative analysis. Most of the analysis techniques used were qualitative. Where quantitative techniques were used, caution was exercised in their interpretation as in no case were the samples of sufficient size and the sampling of sufficient rigour to be considered representative of the populations sampled within 95 per cent confidence, or to meet the criteria of most statistical tests.

The general approach to analysis has been to build a coherent picture of the cases through consideration of all available data. Thus, triangulation across different collection methods combined with due consideration of the limitations of the methods is the best insurance of rigour. Since this research often traded off rigour for breadth, the pictures that emerge are ones 'painted with broad strokes' rather than highly detailed images.

Conceptual framework

To understand the link between universal access policies and on-the-ground experiences, one must first understand the overall context and analyse the ability of the centres themselves to provide for universal access and contribute to development. This conceptual framework attempts to do both. By abstracting from the immediate situation and identifying key actors, processes and factors that exist in different contexts, it also provides a tool for generalizing across different contexts while remaining sensitive to the differences within these contexts. For example, we will see that the Ugandan government's path to universal access (as depicted in Figure 1.1) involved market stimulation whilst the South African government took a more direct approach to implementation through the purpose-specific Universal Service Agency. The final chapter summarizes some of these similarities and differences and from these draws general lessons.

Universal access policy arena and process

The policy arena
The formation and implementation of universal access policy take place in an 'arena' consisting of various actors from the public, private and civil sectors, operating at the local, national and international level, as illustrated in Figure 1.1. The nature of the relationships and influences between all these actors is a factor in understanding differing national experiences in supporting shared access centres, and in identifying possible future interventions to strengthen these initiatives. Figure 1.1 lists some of the generic actors in both South Africa and Uganda, as well as other countries. The specifics of the national and local situations in each country are further detailed in Chapters 2 (South Africa) and 3 (Uganda).

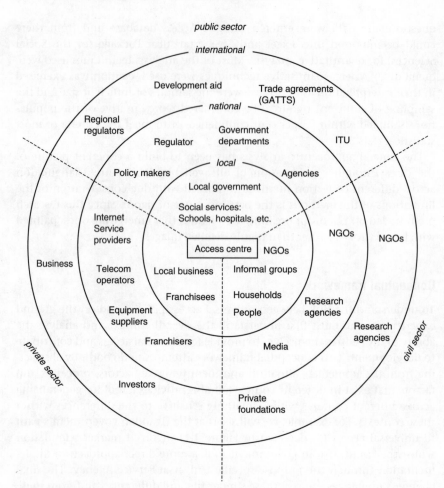

Figure 1.1 Policy arena

The policy formation and review process

General aspects of formulating, implementing and revising universal access policy are shown in Figure 1.2. While this diagram fairly describes the main components in South Africa and Uganda (though not necessarily those in other countries), the emphasis on different means of implementing such policy, the details of both implementation and policy, and the relationship between mechanisms are unique to each country. This is covered in further detail in Chapters 2 and 3. For each national case study, a small number of community case studies are used to focus on the apparent effects of ICT policy and national-level activities upon ICT access within the communities generally, and shared access centres specifically. These are then compared with the original goals of the relevant policies.

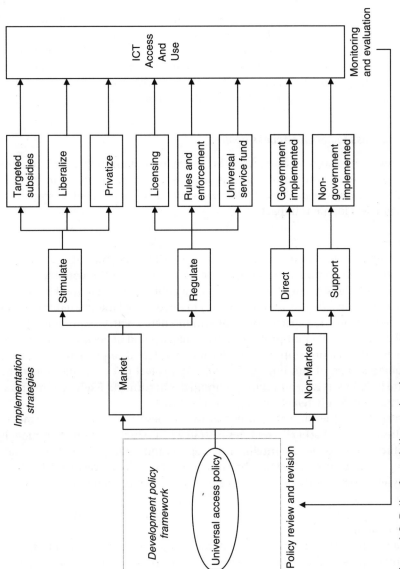

Figure 1.2 Policy formulation and review process

The policy intent behind universal access and service to ICTs occurs within a broader framework of a national development policy. The policy will outline specific implementation mechanisms. Some of these may operate by stimulating and regulating the market. Through licensing telecom operators, governments can impose certain obligations upon them but may also grant them privileges, usually monopoly rights. One obligation is to contribute to a fund that supports access initiatives. Fund resources may be used to subsidize market enterprises or to support government or not-for-profit activities. In addition, non-governmental actors may undertake access-related initiatives or projects independently. The government may wish to support or monitor these initiatives so that they better contribute to overall access targets.

Four success factors for access centres

This research focuses on four factors as criteria for understanding the strengths and limitations of the various models of access centre encountered in the case studies and elsewhere. These factors are scalability, sustainability, reach and use, and development impact.

Scalability

Scalability in the context of this research means the capacity to provide enough access centres to reach the population of an entire country. At the micro-level, the key questions related to scalability are: What are the requirements to set up a centre, and who will actually do it? Market-based initiatives will depend on the capacity of local entrepreneurs, entry requirements and local demand, while non-profit initiatives may depend on local organizational capacity, the relationship between local and external implementers, and an adequate source of funding. Both will require technologies that allow the centre to offer services, and these technologies in turn will likely require infrastructure, especially telephone lines and electricity.

Scalability in the context of universal access doesn't demand that one model fit all situations. Rather, it requires a strategy in which a number of models and broad approaches are intelligently applied and locally adapted across different circumstances, allowing the entire population to be reached. Chapter 4 analyses the scalability of different centre types encountered in the case study areas.

Sustainability

Sustainability, the capacity to continue once begun, is a natural second requirement for access services. From a broad perspective, it doesn't much matter whether the services persist because a single centre persists, or because when that centre fails, another one or more take its place. The continuation of services will normally depend on local demand and the ability of local actors to recognize and respond to that demand. Initial access centres in previously underserved areas are sometimes described as 'market primers' because they are able to expand the local market for ICT services, creating a particular kind of sustainability that may outlast the centre itself (Fuchs, 1997).

Sustainability depends on factors both external and internal to a centre (see Box 1.1). Key external factors include the local market, national market and outside support mechanisms. Internal factors play the ultimate role in determining whether a centre will succeed, since ingenuity and strong management may overcome many obstacles. Yet external factors are important determinants of how tough an environment is and, in most parts of Africa, they add up at least to 'very tough'. All of these factors can be influenced to some degree, and often directly, by national policy. If large numbers of centres are failing within a country, the cause is more likely to be a set of inhibiting external factors than widespread manager incompetence.

Box 1.1 Key factors in sustainability

Internal management and leadership: Vision/champion, capacity, cohesion versus complexity in decision making, level of internal conflict, ability to partner, network, technical know-how, marketing skills, flexibility and quick response.

Internal assets: Building, location, equipment, available services.

Local market: Awareness, income levels, perceived relevance, capacity to use, effective demand, local technical support and equipment supplies.

National market: Operators, available/legal technologies, prices, technical support.

External support mechanisms: Support to address gaps in the local market, including capacity, awareness and ability to pay (ie, via subsidy); support to build managerial and operational capacity (eg, to enhance procurement, maintenance, business and marketing skills); advocacy on behalf of local suppliers and consumers in national policy and market issues; provision of adequate basic infrastructure (electricity, roads, telecommunications).

Access centre reach and use

To provide universal access, access centres should be priced, distributed and designed such that any person can use the target services if he or she so chooses. This means that the price of the service must be within the reach of prospective users' ability to pay, that the users are aware of the service, and that they have the capacity to use it, or feel able to acquire such capacity. The requirements, then, are based on the characteristics of not only the access centre (supply) but also the potential users (demand).

To understand the contextual relevance of ICT use, this research uses a livelihoods approach to explore why some groups of people use ICTs and others do not, how different groups use them, and the likely impacts of such use. By applying livelihoods approaches across the community case studies, we can gain insights into common motivations for using ICTs and common barriers to use, and how current patterns of use are likely to influence development.

Chapter 5 applies this analysis to different types of access centre, generalizing across contexts where possible.

Development impact

The development impact of universal access initiatives will be the sum of the centres themselves, the process of establishing them, the services they provide over time, and the improvements they engender in people's lives. Some access centres may involve complex processes of community mobilization and these processes in turn may have a development impact, irrespective of the centre's performance. Local employment opportunities, the capacity developed by staff, and the money that flows into, out of, and circulates through the community as a result of the centre also have important implications for local economic development. The services themselves are the focus of most expectations for development impact, and this impact will vary according to the kinds of service, who uses them and the nature of the community.

The dual definition of development given earlier provides the basis for analysing development impact based on two questions. How do the centres support relevant national development goals? And what role do they play in increasing (or decreasing) the overall range of choices of local community members? In assessing the latter aspect, this book focuses primarily on local livelihood activities and priorities, as determined through household surveys and census data.

In Chapters 2 and 3, community experiences with ICTs are compared with development goals set out in policy. Chapter 5 focuses on development impact of access centre services, but also of productive opportunities associated with providing services, through the lens of livelihoods.

CHAPTER TWO

Access centres and South Africa's universal access policy

This chapter reviews the role that shared access centres have played in universal access within South Africa, with a specific focus on the policy intent driving universal access.[1] After reviewing the overall ICT policy intent and some of the mechanisms for achieving universal access, it focuses on the experiences of two communities. One is a small rural settlement in Kwa Zulu Natal, north-east of Durban; the other a large township in the Western Cape, south-east of Cape Town, on the same peninsula (see Figure 2.1).

South Africa's experience is interesting. It took a leadership position in Africa, with top-level political commitment to universal access to ICTs. This engagement was driven by the need to redress the legacy of apartheid, and by a

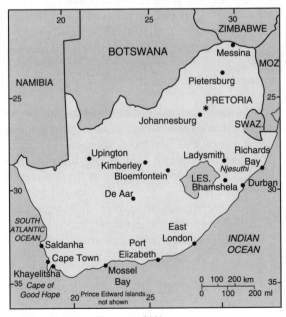

Figure 2.1 Map of South Africa (Source: *CIA*)

strong belief that communication and broad access to information were keys to democracy and development. Moreover, South Africa's government had relatively more money and human capacity than its neighbours to address the task. However, nine years after it began, the results have been largely disappointing. In communities examined here, telephone access has improved greatly, mainly due to cellular phone networks, but rates remain high, and the Internet is barely present and rarely used. Much of this has to do with the market and regulatory environment at the time of research. The incumbent landline telephone operator had a strong monopoly and had kept rates high and connectivity options outside of central business hubs limited (Emdon, 2003). Any process of policy reform is also an opportunity for various actors to pursue their own interests. The number of actors influencing policy within South Africa is large and the process itself has been complex. The situation in South Africa is a reminder that feedback from the experience of implementing universal access initiatives may be a lesser influence on policy revision than political interests.

National optimism and the emergence of universal access policy

Historical context

Universal access policy was developed in the historical context of the end of apartheid and the dawning of a new democratic era in South Africa. This was a time of great optimism and activity. Key leaders, including Nelson Mandela, the first democratically elected president of South Africa, and Thabo Mbeki, his deputy president, and later president, believed that ICTs were important to the development of South Africa. Indeed, they made public statements in high-profile international gatherings to that effect. Their interest converged with increasing global interest in the issue of the information society and the digital divide amongst governments, multilateral agencies and civil society groups.

South African civil society also believed firmly in the importance of information and communication and the tools that could facilitate these. This interest could be considered a legacy of apartheid, since the previous government, through censorship and attempts to control and limit information sharing and communication, had reminded its opposition that these could be effective tools for empowerment.

Finally, the opening of South African markets during the time of transition in the early and mid-1990s represented an interesting and potentially lucrative opportunity for both international and domestic companies. This resulted in great activity and high expectations for ICT-related developments in the country.

South Africa, as a signatory to the WTO GATS in 1998, was committed to telecom sector privatization and liberalization. Global pressure to undertake these reforms was present before this and created some tension within South

Africa. The African National Congress (ANC) had strong ties to the Communist Party and to labour unions, which had also been instrumental in resistance to apartheid. These were opposed to both privatization and liberalization measures since they would reduce sovereignty by selling off state assets to foreign-owned companies and would likely result in heavy job losses within the newly privatized entities.

Development policy framework

The Reconstruction and Development Plan (RDP) base document, released in 1994, was the key framework for the new government's activities and provided the overarching context for the development of the original universal access policy. It referred to ICTs as powerful tools for multi-sector development and emphasized the need to strengthen a diversity of media, including community media, and to encourage the exchange of information between citizens and governments. Within this framework are the Urban Renewal Strategy and the Integrated Sustainable Rural Development Strategy. Both emphasize mobilizing people to become active participants in their own development and tend to focus on infrastructure development, which was lacking in Black-populated areas under the apartheid regime. The Rural Development Strategy aims to create sustainable rural communities capable of attracting and maintaining skilled, knowledgeable people through capacity-building, provision of universally accessible social services, and creation of vibrant local economies (CMS Task Team, 2003).

Universal access policy

The key policies guiding universal access to ICTs are the Telecommunications White Paper of 1996, the Telecommunications Act of 1996 and the Telecommunications Act Amendment of 2001. The 1996 Act set up the regulatory body for telecommunications, which in 2000 merged with the broadcasting regulator to form the Independent Communications Authority of South Africa (ICASA). It also set up the Universal Service Agency as a statutory body without any power to regulate or enforce.

The Universal Service Agency was a uniquely South African invention – no other country had created a separate body devoted to universal service and access. This was evidence of strong national commitment to the issue (Benjamin, 2001a, b). The Agency was mandated to:

- explore and promote innovative ways to promote universal service;
- increase public awareness around the benefits of telecommunications;
- encourage and guide other universal access schemes;
- survey and evaluate the extent of universal service;
- develop and revise definitions and targets for universal access and universal service;

- administer the Universal Service Fund by paying subsidies to assist 'needy persons' in accessing services. This also required developing a definition of what a 'needy person' was.

One problem with the Act is that it did not clearly establish the roles and responsibilities of the Agency and the regulator, and there appeared to be some overlap (Benjamin, 2001a, b). Also lacking in the Act were clear definitions and benchmarks, which the Agency was supposed to establish. However, at the time of this research, seven years after the original act, these definitions were still missing, which seriously undermined any ordered and logical attempt at implementing and monitoring access strategies.

The Telecommunications White Paper of 1996 had defined universal access as 'every person within a 30 minute walk of a public telephone'. The Universal Service Agency, however, also considered itself committed to achieving universal access to advanced telecommunications, including the Internet. No clear strategy for doing this was ever developed. A few discussion papers and public discussions were begun but not concluded in 1998. The process was due to restart in 2003. Sam Galube, who became chief executive officer of the Universal Service Agency in early 2003, defined universal access to ICTs as follows: 'Each and every household having access to telecommunications technology .... This access must be affordable and convenient ... within a 10 to 15 minute walk, a person should have a point where he or she can access a telephone or the Internet'.

The 2001 Telecommunications Amendment established multimedia licenses and laid the groundwork for the granting of licenses to telecom operators in ten (out of 27 identified) underserved rural areas. These licenses were being put out to tender at the time of research and therefore were not yet in effect. The Universal Service Agency's mandate was expanded to include the administration of subsidies of about R10 million to these licensees (Lewis, 2003). A Convergence Act was passed in December 2003, after field research.

Besides these core documents, numerous task teams and advisory bodies have produced reports for the Presidency, Cabinet and various sectoral departments. The Department of Education, the Department of Trade and Investment, and the Department of Arts, Culture, Science and Technology have been actively involved in issues surrounding ICT access provision and related capacity building, and each has its own projects and related policy. The large number of activities within various sectors of government poses a challenge for co-ordination. Fruitful engagement with non-governmental actors has been a further challenge. Recent policy formulations attempt to address this by acknowledging existing bodies and their efforts and emphasizing the need for partnership with them.

Table 2.1, which draws on the over-arching documents and public statements from leading government officials, summarizes the key policy intentions regarding universal access in South Africa.

Table 2.1 Summary of South African national policy goals related to universal access

Goal	Evidence/where stated (not exhaustive)	Implementation mechanisms
Universal access to telecom (and broadcasting) service, redressing historic inequities	Telecommunications Act 1996, revised Act 2001, mandate of Parliamentary Portfolio Committee on Communications	Regulatory authority (ICASA), licensing agreements, Universal Service Agency, Universal Service Fund
Delivery of government services including social services (e-health, e-education, etc.)	Eg, Department of Education White Paper on E-education (2003), research commissioned by PNC ISAD	Various government departments, GCIS, PNC ISAD, Universal Service Agency under 2001 mandate
Improvement of communication between government and citizens	RDP 1994	Various government departments co-ordinated by GCIS (own services)
Democratization, diversification of media and expression	RDP 1994, MDDA Act 2002	DoC, MDDA, Department of Arts and Culture
Supporting local development through information provision	Mandates of DoC, Universal Service Agency	DoC, GCIS, Universal Service Agency, various government departments
Supporting SMMEs and job development	ICT Economic Empowerment Charter (2004 in progress)	Department of Trade and Industry, ISETT SETA, Universal Service Agency under 2001 mandate
Creating a South African information society	Presidential speeches, eg 2001 State of the Nation address, Telecommunications White Paper 1996	Various: DoC, Universal Service Agency, Department of Arts and Culture, presidential task forces

Implementing universal access

South Africa has a large, somewhat confusing array of government pro-grammes and non-governmental programmes and projects aimed at providing access to ICTs, media and related services. In addition, its telecom sector has gone through a number of tumultuous regulatory changes. The market situation was likely to change as a new convergence bill came into effect and a second national telecom operator finally became licensed (2003–2004). The market was one of the most visible influences across the whole telecommunications sector in South Africa, also limiting governmental and not-for-profit efforts, which by law depended upon licensed operators for basic services. For example, wireless technology that expanded beyond private property for private use was illegal if implemented by any other than the incumbent telecom operator.

The main mechanisms by which ICT access was provided in South Africa are:

- the market: cellular telephone, computer, fax, Internet;
- licensed telecom operators, who were given both rights and obligations to fulfil under their licenses, regulated by the Independent Communications Authority of South Africa (ICASA);
- licensed radio and community radio stations, the latter also receiving support from the Department of Communications and the non-governmental National Community Radio Forum;
- the Universal Service Agency's telecentre programme;
- multipurpose Community Centres of the Government Communication and Information System (GCIS);
- a variety of largely independent, uncoordinated, not-for-profit initiatives, often affiliated with libraries, community centres and local NGOs;
- school computer labs.

The most prominent of these (based on the findings in the community case studies) are described below.

Licensing obligations, regulation and the market structure

'Telkom is very expensive. Sometimes we get phone bills that are too high; but you can't argue with Telkom because while they have you cut off from services your business is dwindling. Soon they will introduce ADSL here, which has a flat rate. That will make a big difference'.

Business owner, Randburg

Telkom, the former South African parastatal, was given a five-year monopoly for fixed-line service plus a one-year extension that officially ended in early 2002. Because no second national operator had been identified, the monopoly was still in effect during the time of research in 2003. Telkom, whose monopoly extended to most voice- and data-transmission services, was required by its license to provide 2.7 million new lines, including 1.7 million in under-serviced areas and 120,000 pay phones, plus contribute about R40 million to the Universal Service Fund (USF).[2]

Sentech, the major broadcasting signal distributor in the country, was given a multimedia license, including the right to license others, in May 2002. This created partial competition in data transmission and broadcasting, but the effects were limited. For example, various options for accessing the Internet such as wireless broadband and very small aperture terminal (VSAT) remained expensive and their deployment was limited.

The cellular market included three operators: Vodacom (with 65 per cent market share), Mobile Telephony Networks (MTN) and Cell C. MTN and

Vodacom were granted licenses in 1993 and Cell C was licensed in 2001. By October 2003, South Africa had 15 million cellular users and the fourth fastest growing market in the world (Burger, 2004). Low-income groups, constituting the vast majority of people, use pre-paid card services as these do not require up-front payments or a contract stipulating income conditions.

Community service obligations for the cellular operators were set at 22,000 for Vodacom, 7,500 for MTN and 52,000 for Cell C. These obligations required each operator to set up the requisite number of public phones in 'under-serviced areas'. Because densely populated townships were considered underserved, but were relatively easier and more profitable to cover than rural areas, operators focused most of their efforts in these areas (Benjamin, 2001a, b). Community service obligations were also, retrospectively, set very low given how lucrative the market turned out to be. Despite this limitation, Vodacom's approach to community service was generally seen as very success-ful and was later emulated by the other operators.

All licensed operators have been required to pay into a Universal Service Fund (USF) since the Telecommunications Act was passed in 1996. Until 2001, contributions to the fund were capped at a total of R20 million a year, which was very low compared with the revenue generated (much less than 0.1 per cent of turnover) and the requirements of other countries. In 2003, ICASA set the actual required contribution at 0.2 per cent for each licensee, still low but a substantial increase (Msimang, 2003).

Access centres and cybercafés, whether for-profit or NGO, were not specially licensed. However, they were profoundly affected by the market and licensing conditions of the large infrastructure providers as they were all required to buy their services from the licensed telecom operators.

The Universal Service Agency's telecentre programme

'When [the Universal Service Agency] started off, it was about putting boxes in communities. Now we see it's about what those boxes deliver – services such as school subjects, computer literacy and e-health. We will partner with companies to deliver these'.

Dennis Mamela, Universal Service Agency

The Universal Service Agency was established by the 1996 Telecommunica-tions Act to explore and promote creative methods of universal service and access. The original Act established it for five years. During this initial period, the Agency devoted most of its energies to supporting the creation of tele-centres across the country, paid through the Universal Service Fund (USF). It invited applications from NGOs and community organizations. Those selected were given equipment and basic training; from that point, they were supposed to run the telecentres as self-sustaining entities.

While its original target was hundreds of such telecentres, the Agency developed only 65, which experienced a variety of problems almost immedi-ately. By 2001, one-third were not functioning, half did not have phones, and only a few had access to the Internet (Benjamin, 2001a, b). By 2003, the situation had deteriorated.

The Universal Service Agency itself was hampered by a poor reputation, internal management difficulties and poor relationships with other key organizations (Benjamin, 2001a, b). It was unable to get crucial information from operators that would allow it to map telecom provision and thus effectively monitor progress. It never developed definitions, indicators or benchmarks for any key terms that would allow it to strategize effectively. In addition, the Agency, partly in response to political pressure, moved quickly towards an attempt at massive rollout of telecentres without proper piloting of the concept.

In 2001, the Universal Service Agency's mandate was renewed and expanded. And in 2003 it underwent internal restructuring and renewal, including an audit of the remains of the existing telecentres. The Agency is now working more closely with the Department of Communications and the GCIS (see below), and is pursuing partnerships with other actors including Sentech.

The GCIS Multipurpose Community Centre (MPCC) Programme

'MPCCs need to be more than just government service centres'.
    Holly Luton-Nel, Director of Aleksan Kopano Resource Centre, Alexandra

The Government Communication and Information System (GCIS) began operations in 1998 with the goal of increasing the accessibility of government information and services and improving communication between the government and the people. The focus has been on establishing Multipurpose Community Centres (MPCCs), with an initial goal of 65 MPCCs: one in each district municipality. The basic concept is that of a shared facility housing at least six government departmental offices plus a telecentre (established by the Universal Service Agency). The centre may also house community groups and facilities, including community radio stations. The departmental offices provide relevant information and services to the local community, including training and outreach activities. The telecentre provides telecommunication services to other MPCC tenants, allowing it to provide public services at reasonable cost and at the same time contributing to the telecentre's sustainability. At the time of research, about 30 MPCCs had been established across the country and others were to be established in the following year.

Non-governmental initiatives
Various external, national and local non-governmental organizations have launched their own initiatives. These tend to focus on each organization's own implementation activities, although there is some partnering and information sharing. Links between these initiatives and government initiatives have generally been weak. Thus, government policy and efforts have not acted

as a co-ordinating or unifying force across these independent initiatives, and shared learning has also been limited.

Access achieved? Review of the case in two communities

Bhamshela

'In 1998, people didn't understand what a computer was or what it could do. The person who stole our school computer buried it'.

High school principal, Bhamshela

'In the past, we'd identify a CBO [community-based organization], donate computers, train them and then the organization would be responsible for the running costs, including paying the staff. But the centres were serving poor communities so business isn't attractive. They were unable to pay salaries and the staff would end up taking money from the telecentre'.

Universal Service Agency co-ordinator, Kwa Zulu Natal

Bhamshela is a densely populated rural community and trading centre, less than two hours' drive from Durban. Until 1998, when a telecentre was established, there were few telephones. The telecentre predated all other public ICT services and was very popular. While it offered a full range of services, including computer and Internet rental, only telephone and, to a lesser degree, photocopying services were used. This was partially due to the lack of expertise of the staff in using these, and partially to lack of awareness and demand on the part of potential clients.

The telecentre ceased phone, fax and Internet services in 2000 after experiencing problems with the telephone company, resulting in a large bill it could not pay, exacerbated by problems with its phone metering system. Local demand for phone services persisted, though, and a Vodacom shop and Telkom container were opened the following year, the latter by a former staff member of the telecentre. Meanwhile, the telecentre further developed its staff's expertise in computers and began to offer basic training courses. There was demand for such courses due to the high priority many people placed on finding employment and the perceived importance of computer skills in formal-sector jobs. Many students defaulted on their tuition payments because they could not afford them. A few found jobs.

A GCIS-implemented MPCC officially opened in 2002 and began operations in 2003. It had depended on being able to receive ICT services from the telecentre, but because the telecentre could not deliver, the Village Bank began to offer fax services and later intended to launch public Internet services.

The full impact of the telecentre is difficult to assess with certainty since phone penetration, especially in the cellular market, increased quickly across the country over the same period. For many local people, the telecentre was the first place they made a phone call. Many people knew it had computers (although many didn't), and the students taking the courses had often done so on the recommendation of others. It is also clear that the telecentre motivated other valued community development processes, including the application for the MPCC and establishment of a free bi-weekly health clinic.

Lessons

It is not clear that the telecentre, with its bundle of services – some appreciated, some not – was the best model for this community. Arguably, it should have been either explicitly wider in scope to formally include facilitation of community development in its mandate (which would require funding other than from user fees), or else more focused on locally relevant ICT services. A needs assessment could easily have identified both the demand for voice telephony and for job-related computer skills (in fact, any job skills). These could have been delivered more efficiently and effectively via two separate means. For example, the Vodacom container could offer lower prices than the telecentre while generating a better profit margin for its owner. Part of the reason is that uneconomical services offered by the telecentre must be subsidized by services in demand such as phone, which affects pricing. Since phone is the service most in demand, and cost is a limiting factor in accessibility, such a strategy does not appear to serve the public interest.

The computer classes were not part of the telecentre's original service mix because the staff did not then have the capacity to deliver them. Resorting to this training service was as much a survival strategy as a service to the community. In this sense, they relied too heavily on phone services while they were able to do so and demand was high. Again, this does not appear to serve the community in the best manner. A clear needs assessment and targeting of resources where necessary – for example, on staff training and a set of appropriate textbooks – would be a better use of money than setting up a clearly unsustainable, unaffordable Internet connection.

This leads to a third major lesson that can be drawn from Bhamshela: offering Internet services where they are too expensive to be sustained and where local demand is absent only creates a further strain on the overall viability of the telecentre. At very best, it is a symbolic gesture and a service to a handful of people. There are two basic options: introduce a different pricing structure so that local demand can be accommodated and increased over time (implying subsidization over the medium term), or focus the organization's efforts in areas likely to yield better results.

Finally, the MPCC appears to offer great potential for developing the area, and ICTs can clearly enhance its services. The situation in Bhamshela has created an inadvertent but interesting example of a strategy whereby the government has indirectly supported access to ICTs by providing an anchor market that stimulated a private entity (the Village Bank) to provide the actual services.

Khayelitsha

'Lack of knowledge of computer skills is one reason for the high unemployment in Khayelitsha, and may also be one reason why the Public Information Terminal is underutilized'.

Director, KERIC, Khayelitsha

'In Guguletu, computers and the Internet are more in demand [compared
with Khayelitsha] because of multiracial schools. The public libraries there
are full and an Internet café just opened three or four months ago. In
Khayelitsha, nothing has really changed in the last three years except
competition is getting higher [for phone services]'.

Local phone shop owner, Khayelitsha

Photo 2.1 Khayelitsha's shopping centre

Khayelitsha is a fast growing, densely populated, predominantly Xhosa
township with a population of about one million people. It is located on the
Cape Flats about 30 kilometres from Cape Town. Despite its large population,
it sustains only limited economic activity, with most residents shopping and
working elsewhere. It contains a combination of informal and formal settle-
ments with varying levels of infrastructure. Both public phones and private
cell phones are widely available and phones are used by over 90 per cent of the
population. This situation is markedly different from 1996, when only about
half had made at least one phone call. However, the use of other ICTs remains
quite limited, mainly due to cost and lack of capacity.

The market in Khayelitsha sustains many phone shops initiated in part due to the license obligations of the major operators, plus some independent phone shops. It also has at least one large computer training centre and a couple of other businesses that offer non-phone ICT services in addition to their core business (a phone shop and a driving school respectively). These include value-added services such as typing CVs, letters and funeral programmes. Most of these are centred in and around one of Khayelitsha's few shopping areas, the Sanlam Centre shown in Photo 2.1.

The non-profit sector in Khayelitsha offers a similar range of ICTs. In addition, one library offers Internet access. Although some entities have tried to do so, public Internet provision cannot be sustained via the market due to the large gap between the cost of providing it and people's ability (and willingness) to pay. There need to be cheaper ways to offer Internet access. The high prices can be attributed largely to Telkom's monopoly in effect at the time of research. Given the experiences at the library, market prices would have to be substantially lower before people would use Internet services. This implies that even in a fair and competitive market, many people would not consider the Internet an affordable and justifiable expense if they must pay by the minute or hour.

Lessons

In Khayelitsha, providing phone service can be done within the market, while providing Internet service cannot. Other ICT-related services such as CV typing and faxing are barely sustainable within the market, and generally have to be combined with other services, whether ICT-related or not.

Finding work is the first priority of most people within Khayelitsha, and a gap between skill levels and the requirements of formal-sector jobs is generally understood as a large contributing factor to unemployment. Yet there are relatively few services in place to address this. Megabro is a private-sector training organization while Learn to Earn, a Christian NGO, provides computer training with a focus on creating self-employment opportunities.

The failure of the market to address the local need for ICT training and to harness ICTs for local entrepreneurial opportunities is attributable only in part to the high costs of telecom services. It is also reflects historical economic neglect within Khayelitsha and the tendency of residents to conduct business outside the township. This in turn implies that strategies to provide access to ICTs should be integrated with, or complementary to, other initiatives to stimulate local economic development. Targeted training and capital loans for current and potential entrepreneurs are two ways to stimulate such activity. However, non-phone ICTs are high-risk ventures and appear to be most successful when introduced into existing enterprises. They should be encouraged only where careful market research shows they are viable.

Comparing the South African cases with access-related development objectives

Table 2.2 summarizes the study findings as they pertain to relevant national objectives, which were compiled from policy documents and public statements. In some instances, 'progress' is a rather subjective term. For example, there are no indicators or benchmarks to give precision to what a South African information society would look like, especially in townships and rural areas.

Table 2.2 Development goals versus actual progress

Goal	Progress in Case Study Communities
Universal access to telecom service, redressing historic inequities	*Khayelitsha*
	Good progress in telephone: combination of public and private phone services accommodates over 90% of population.
	Internet penetration and use are extremely limited.
	Bhamshela
	Phone accessibility has improved greatly, although cost and geography remain barriers.
	Access to other ICTs remains low because of limited awareness, high costs and no clear relevance to most people's needs.
Delivery of government services including social services such as e-health and e-education	*Khayelitsha*
	Government social services have improved greatly in Khayelitsha, although ICTs are not an important delivery mechanism. There is one public information terminal (PIT), reportedly underutilized.[1]
	Bhamshela
	The MPCC is delivering a range of locally relevant government services, although ICT use is not a central component of delivery at this time.
Improvement of communication between government and citizens	*Khayelitsha*
	This has not occurred via the ICT services reviewed in this research. If effective, the PIT provides a one-way (government-to-people) flow of information. No evident demand for this from the community.
	Bhamshela
	The MPCC was too new to assess but there is potential for ICTs to support its operations.
Democratization, diversification of media and expression	Khayelitsha
	Radio Zibonele, the local community radio station, is widely listened to and appreciated. Some computer secretarial services offered limited desktop publishing capacity; this was used to produce items such as funeral and wedding programmes.
	Bhamshela
	None of the ICT-related facilities in Bhamshela is related to media or local information production.

Table 2.2 – *continued*

Goal	Progress in Case Study Communities
Supporting local development through information provision	*Khayelitsha* NGOs such as KERIC were doing this in the context of a larger advocacy role. However, this was not linked to their ICT services. Other NGO activities such as the Lovelife social marketing campaign provided information targeted at social change in the face of AIDS. Links between these and ICT access were not evident in the case studies. *Bhamshela* The MPCC has this as part of its mandate.
Supporting SMMEs and job development	*Khayelitsha* Public phone services were a strong area of entrepreneurial activity. Selling connection time and cell phone accessories, as well as repairing cell phones, radios and TVs, were also popular as small enterprises. Since there are relatively few job opportunities within Khayelitsha, the amount of activity in this sector appeared significant. *Bhamshela* The Department of Labour has a presence at the MPCC and offers related services. The Vodacom container provided two local jobs, and the computer training at the telecentre had helped at least three local people find work. Paid and volunteer positions at the telecentre also allowed several people to gain experience and improve their employability, although the centre itself proved a poor employer. Other ICT services such as fax and photocopying nominally offer business support services, although there was little local demand for these.
Creating a South African information society	*Khayelitsha* Most residents use radio and phone but have little awareness of or opportunity to access digital ICTs such as computers and the Internet. People see these digital ICTs as important for increasing their job prospects. *Bhamshela* Phone services have eased communication in and out of Bhamshela considerably. Other ICTs are not part of the daily life of the vast majority. Interest in computers is widespread despite low awareness, both for employment opportunities and because people understand them as a part of the modern world.

[1] The South African Department of Communications was involved in implementing Public Internet Terminals (PITs) from about 1998. These were Internet kiosks installed in some post offices in disadvantaged areas. Many did not actually have Internet connectivity and simply provided information from a CD-ROM (Benjamin, 2001a, b).

Learning from South Africa's experience

South Africa's situation is in many ways unique, linked to its history and the fact that it remains, after Brazil, the most inequitable country in the world in terms of wealth distribution. Despite the great optimism unleashed by the coming of democracy, the legacy of its past and the great economic division that still characterizes the population influence the behaviour of both the large dominant telecom agencies, especially Telkom, and poor, historically disadvantaged communities.

Because there is such great wealth in the large cities – the Province of Gauteng, home to Johannesburg and Pretoria, for example, generates 9 per cent of the GDP of all Africa – and the infrastructure is much better there, large telecom companies may be tempted to focus on urban markets, avoiding poor townships and rural areas. Pricing is not geared towards these latter areas and certain services, such as ADSL (asymmetric data subscriber line), may not be available much beyond the major urban centres.

South Africa reportedly has one of the lowest levels of entrepreneurship in the world. Within both townships and rural communities, there is relatively little entrepreneurial interest or risk taking beyond very simple and proven business models. Telephone services have become widely recognized and coveted as small business ventures, especially where franchises are available. Relatively few entrepreneurs have explored other forms of ICT service provision; and those who have often have not done well. Because of high overhead costs and low profit margins, most small businesses offering ICT services do so in addition to another pre-existing business activity (eg, a hair salon, traditional healer, or driving school). Such hybrid businesses were quite limited in number in the two case communities studied. They were more visible, however, in large, high-traffic urban areas such as Johannesburg, Randburg, Soweto and even in the business districts of mid-sized towns.

Although Bhamshela is rural and Khayelitsha is urban, the two areas were observed to have similar access and use rates, especially for computers and the Internet. This was mainly due to the lack of Internet access in both cases and, further, because the Bhamshela telecentre had exposed some of the population to computers and, briefly, the Internet. People throughout the country who had experience dealing with Internet access issues commonly identified the high cost, and to a lesser degree the low quality, of Internet connectivity as factors that had severely limited or blocked their efforts to offer and use it in communities and schools. This in turn was generally attributed to the failure to liberalize the telecom market and effectively regulate price increases, which were unjustifiable in face of the profits being made (Emdon, 2003).

The South African government took a direct role in implementing access centres, through the activities of several departments and the Universal Service Agency. This experience was generally not very successful. Neither of the Agency's telecentres in the case studies was well used and both had severe financial problems. Based on anecdotal and documented experiences of other telecentres, their experiences were typical. A major reason for this failure was simply that the larger issue of high cost was not addressed. Many of these centres reported receiving poor service and having to pay high costs with little profit in return. On the implementation side, roles and strategies were not well defined, and there were never clear benchmarks to aim for. The Agency, reporting to the Department of Communications, was sometimes under political pressure to act before it was ready to, at the same time suffering from various problems of internal capacity that rendered it relatively ineffective and gave it a poor reputation (Benjamin, 2001a, b).

The Agency might have been a more effective champion of universal access in South Africa had it been more closely aligned with, and perhaps even a part of, the regulator. As such, it would have had greater power vis-à-vis the telecom operators. It also required greater independence and internal capacity so that it might play an advocacy role on behalf of telecommunications consumers and small service-providing entrepreneurs. Given the general market condition, this role was more urgent than that of implementer. Luckily, a number of NGOs have taken on this role to some degree. These include Sangonet, the LINK Centre at the University of the Witswatersrand, and bridges.org. By engaging in independent and public research and monitoring, these organizations provided an important counterbalance to the vested interests of the large telecom companies that otherwise dominate the field. The South African media, especially the business media, have also given a fair amount of coverage to telecom issues and have voiced criticism of the disadvantages of monopolies and the government's slow movement towards liberalization.

South Africa has made significant progress towards universal telephone access, although not in the way foreseen in policy. In particular, cellular technology and the huge uptake of cellular phones, both through personal ownership and phone shops, were not anticipated in the policy. Since many people had their phone service disconnected (an estimated two million lines), the issue of landline service has not been one of installation and availability, but of affordability. Ironically, this has been the case with phone shops and telecentres as well – entrepreneurs cannot afford the narrow profit margins and poor service from Telkom; they much prefer Vodacom or one of the other cellular operators. Telkom had recently adopted pay-as-you-go landline services in an attempt to imitate the success of this payment system for cellular phones. The effects of this were not yet clear by the time of study, but some people were reconnecting their landline phones with this service. Even the Khayelitsha telecentre was considering using it.

The Vodacom phone shop model seemed particularly popular with both entrepreneurs and customers. The service was reliable and, for the shop owners, the great advantage of Vodacom's preferential rates, compared with arrangements with other phone shops, was that they saw gross profit margins of about one-third. In return, they helped to meet Vodacom's universal service obligations and advertised the Vodacom brand amongst potential new cell phone users. The preferential in-network rates also meant that customers tended to maintain loyalty to the Vodacom network. After about four years, the Vodacom phone shop model, although undertaken to meet its licensing obligations, was not costing the company anything.

In South Africa, the first government priority would appear to be the true liberalization of the telecommunications market. The second is to ensure that smaller South African companies, entrepreneurs and potential employees from disadvantaged groups share in the wealth generated by the rapid growth of the telecom sector. Thus, for poor communities the benefits of sectoral growth should outweigh the costs. Two important movements in that direction were the ISETT SETA, which supported training within the ICT sector amongst historically disadvantaged populations, and the ICT Black Empowerment Charter, which was still in draft form in 2003 (but due to come into effect in March 2005). The intention of the latter is to give preference to black-owned businesses under certain circumstances. Another area of priority is putting computers in schools and ensuring that teachers and students can use them. For most South Africans, the acquisition of practical job skills is an absolute priority. The provision of computer training in public schools is one of the best ways to achieve this.

Finally, given the great inequalities within South African society, an organization devoted to universal service could serve an important role in safeguarding public interest and in strengthening the capacity of people to create access within their own communities. Here, small business and not-for-profit organizations have an important role to play. At the time of this research and before, the Universal Service Agency appeared to be a weak and ineffective institution. However, if it were able to step back from implementation and focus on monitoring, research, advocacy and capacity building, including provision of small grants and loans to entrepreneurs from historically disadvantaged areas, then it might be able to further the cause for which it was created.

CHAPTER THREE
Uganda's experience with shared access centres and universal access policy

Overview: Universal access policy and implementation

This chapter examines policies concerned with universal access in Uganda, the mechanisms for implementing them, and the on-the-ground experiences across three different cases. The case study communities are Lira municipality (surrounding the town of Lira as shown in Figure 3.1), Nabweru subcounty (slightly north of Kampala), and Kabale district, including Kabale town and the rural subcounties of Rubaya and Bukinda.

Uganda is often held up as an example of success, for other countries to take note. It was able to partially liberalize its telecom market quite early and,

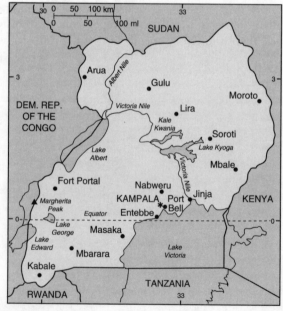

Figure 3.1 Map of Uganda (Source: *CIA*)

against common convention, it introduced a second major operator before it had privatized the incumbent. Somewhat ironically, part of the reason for its relatively smooth transition was that the incumbent operator had been losing money – which meant there was little incentive for the government to try and maintain control. Three cellular operators, especially the most popular and widespread operator, MTN, have provided cellular coverage to the majority of the country's population. However, as most citizens are poor and rural, the tariffs and associated expenses of owning a phone are still beyond them. A large urban-rural divide is strikingly evident, not just between Kampala and the rest of the country, but between small towns and surrounding villages. This suggests that, despite significant progress, universal access to either phone or the Internet is unlikely to be fully achieved by the market alone.

In towns, entrepreneurial activity, not just in phone services, but also in other ICT-related services, was prevalent. The success of these businesses appeared quite mixed. Profit margins were often low and sometimes negative. As in South Africa, phone services had a ready demand and seemed to be much safer and often more profitable business ventures, although competition was high. In rural areas, information services are unlikely to spark spontaneous market demand despite a perceived need for them, especially regarding market information for farm produce. Integrating such services into other rural development and extension efforts seems the most promising way forward, and a number of projects and efforts in this direction have already been made – specifically the new agricultural extension programme (National Agricultural Advisory Service) and a climate-information programme called RANET, delivered through the Department of Meteorology.

Universal access and policy intent in Uganda

Historical context

Uganda has 26.4 million people (2004 estimate), 86 per cent of whom live in rural areas. Its history following transition from British colonial to independent rule under Milton Obote and Idi Amin was one of oppression. Hundreds of thousands of people were killed, many more were exiled, and the country suffered extreme economic and institutional instability and decline.

In 1986, Yoweri Museveni seized power and established a no-party democratic system. He instigated major economic and social reforms in line with the emerging theories and practices of major multilateral institutions. Economic reform, consistent with thinking at the International Monetary Fund (IMF) and the World Bank, focused on liberalization, privatization and reducing the role of the government in the market to one of regulator and enabler. Uganda's GDP grew by an average of 7 per cent a year from 1990 to 2000. Many exiles, well educated and optimistic, came home. Keen to help the new Uganda reform itself, donor agencies and foreign governments provided large amounts of foreign aid and loans.

Uganda has a decentralized government system composed of five levels. It is broken into regions and then districts which are further broken into counties, sub-counties and parishes. Most local government activity occurs at the district and subcounty level. All development has been guided by overarching frameworks, namely Vision 2025 and the Poverty Eradication Action Plan.

National leaders identified ICTs as a potential tool to support their national development objectives. In 1997, President Museveni attended the Global Knowledge Conference in Toronto, Canada, and requested the global community to assist Uganda in developing ICT capacities that could improve the lives of its rural and disadvantaged communities. A number of other ministers and high-level politicians championed the cause of ICTs as potential tools for national development (Ofir, 2002).

The telecom sector was privatized and partially liberalized from 1995 to 2000. During this period, some ICT-focused pilot projects were implemented by donor agencies. The most noteworthy were the Nakaseke multipurpose community telecentre, established in 1998 by a partnership between ITU, UNESCO and IDRC, and the Nabweru and Buwama telecentres, established in 1999 by IDRC. In 1999, World Links for Development, sponsored by the World Bank, began to implement school-based telecentres in partnership with Schoolnet Uganda.

Before liberalization in 1996, Uganda Post and Telecom was the parastatal responsible for providing telecommunications service in the country. Teledensity was low and concentrated in Kampala. The quality of service and speed of delivery were also low while costs were high. As part of the liberalization, the Communications Act of 1997 established an independent regulator, the Uganda Communications Commission (UCC). Part of its mandate was to establish and administer a Rural Communications Development Fund (RCDF). This, along with licensing agreements and, later, full and effective liberalization, was to be the main tool for ensuring universal access to communications services.

Development policy framework

Ugandan national priorities and strategies have been effectively guided by a number of national development policy frameworks. The 1997 Poverty Eradication Action Plan (PEAP) identifies priority areas and strategies. Progress is reviewed regularly through a Participatory Poverty Assessment (PPA) which includes community-level workshops. Carried out under the Ministry of Finance, Planning and Economic Development, the first PPA resulted in the revision of the PEAP in 2000.

The revised PEAP defines four major goals: fast, sustainable economic growth and structural transformation, good governance and security, increased ability of the poor to raise their incomes, and increased quality of life for the poor. It emphasizes the role of increased market liberalization and private sector development as an indirect strategy to reduce poverty.

Cross-sectoral objectives relevant to ICT-access issues include better public health information, sensitivity to gender issues, and addressing wide geographic disparities, especially in infrastructure. Infrastructure is also noted as important for facilitating private sector development in Uganda, which is in turn a key strategy, although indirect, for reducing poverty. Universal primary education and the Plan for Modernization of Agriculture are two major initiatives aimed at reducing poverty and stimulating development.

The PPAs suggest priorities for poverty reduction and provide feedback on the success of policy implementation. In the most recent PPA, ill health was seen as a primary cause of poverty, followed by limited access to land, lack of markets, high taxes, deaths in the family, alcoholism, polygamy, large families, and lack of women's control over productive assets. In the north and west of Uganda, insurgency and insecurity leading to displacement were identified as key causes of poverty. Access to education, skills, gainful employment (or multiple income sources), and start-up capital were all viewed as key factors in moving out of poverty. Lack of empowerment and knowledge were identified as defining elements of poverty.

Universal access policy

Three policy documents, central to understanding universal access policy in Uganda, are briefly described below.

Uganda Communications Act, 1997
The Act establishes the UCC and the Rural Communications Development Fund (RCDF). It emphasizes that the communication sector should have as little government interference as possible, including direct and indirect subsidies.

Rural Communications Development Policy (RCDP), July 2001
This is the key policy document on universal access in Uganda. Produced by the UCC, it links access objectives to development issues, and provides clear detail on implementation strategies, benchmarks and indicators. Various multilateral institutions, such as the ITU, have showcased this policy as an excellent example of an access policy. Implementation was under way but in the early stages at the time of this research.

The policy explains how the UCC intends to administer the RCDF. It aims to ensure that rural residents have reasonable and affordable access to communications services. Specific objectives are described in Table 3.1.

'Rural' is not defined within the policy. In Uganda, Kampala has a far greater proportion of ICT-related services than any other area in the country, and 'rural' is often used to refer to any place outside it. The UCC uses the political demarcation of the country into regions, districts, counties and subcounties as a basis for planning equitably disbursed services. District capitals, therefore, are the first targets for expanded services and act as hubs for surrounding areas.

As benchmarks are reached, the UCC targets counties, subcounties and eventually parishes.

National Information and Communication Technology Policy, July 2002

The Uganda National Council for Science and Technology (UNCST) spear-headed this policy. It was developed through extensive stakeholder con-sultation, with input from government, the private sector, civil society and donor agencies beginning in 1999. It affirms the government's commitment to ICTs as major tools in Uganda's development, serves to unite otherwise fragmented efforts, both by governmental and non-governmental parties, and creates clear links between ICT activities and overall national development objectives and strategies. While the RCDP more directly addresses universal access issues, what this policy adds is a greater emphasis on content and on the flow of existing information. It puts ICTs into the larger framework of communication systems, including both traditional communication systems and mass media. It also sets up a co-ordinating body so that all ICT-related initiatives, both within government and those instituted by donor agencies and NGOs, work together and are in line with the policy's national objectives.

The policy attempts to strike a balance between content and infrastructure. It has three foci: 'information as a resource for development, mechanisms for accessing information, and ICT as an industry, including e-business, software development and manufacturing' (Government of Uganda, 2002). While infrastructure is not sectoral, information generally is, and the policy includes health, education, agriculture, energy, environment, business, science and technology, and other areas as specific sectors. Taken together, the RCDP and the National ICT Policy identify eight broad areas in which ICT is expected to contribute to national development, as shown in Table 3.2.

Without the implementation plan, the policy is somewhat abstract. It was still being debated in Cabinet at the time of field research, although an imple-mentation strategy was simultaneously under development. This policy is fully compatible with and cognizant of existing communication policies, including the RCDP, which fit as components within this framework. In view of the early stage of implementation at the time of the research, the policy had had few practical effects. Nevertheless, it did suggest that the government was well co-ordinated on the issue of ICTs and committed to their use as develop-ment tools. Having such a policy also allows for greater government control over international development projects using ICTs within Uganda.

Sectoral and other policies

In addition to the core policies described above, there are a number of sectoral policies in place. Again, the ICT policy is intended to act as a co-ordinating framework for these other initiatives, and the 2000-revised PEAP is also an important co-ordinating framework. The following are among the various sectoral ICT initiatives by government bodies:

- The Ministry of Education has developed a technology-enhanced education policy.
- The Meteorology Department has participated in an international initiative (RANET) that uses Worldspace technology to transmit seasonal forecasts to rural farmers, especially in drought-prone areas.
- The Ministry of Local Government, sponsored by the World Bank, is implementing a database across the districts that will allow for greater tracking and transparency of local facilities and activities. This will eventually be networked for better information flow between local and national government and across local levels.

While many sectoral activities are not concerned with universal access per se, the decentralized nature of government and the largely rural population necessitate a decentralized ICT infrastructure. Thus, the activities of the UCC to encourage this are highly relevant, just as innovative efforts within sectors (like the health 'personal data assistant' project) to achieve far-reaching connectivity in a low-density rural environment are of potential significance for other access initiatives. Those initiatives concerned with reaching farmers especially would directly benefit from widely accessible public access points in their own strategies.

Other related policies and actions of note include the scrapping of import tax on ICT-related hardware (computers, etc.) as an incentive for ICT adoption in Uganda, and the 1992 liberalization of broadcasting. Broadcasting is classified as separate from point-to-point communications and is licensed through the Broadcasting Council, outside the mandate of the UCC. Radio Uganda, the state-owned station, broadcasts countrywide in 28 languages. According to a 1998 field survey, it is the major source of information in Uganda (Government of Uganda, 2002). Liberalization has resulted in an explosion of FM stations across Uganda – 119 were operating at the time of research. Thus, most districts now have at least one such station broadcasting over a more limited range in local languages. Only about four community radio stations exist, primarily because licenses are almost impossible to obtain. FM stations have often linked with, promoted, used and sometimes started Internet cafés in mid-sized Ugandan towns, including Mbale, Lira and Kabale

Implementing universal access

'It costs us US$300 a month for the connection, compared to $70 a month for the same service in Kampala. We break even and make money on other services. The equipment is 30 per cent subsidized by UCC'.

Joseph Elotu, cybercafé owner in Mbale

Uganda's national strategy for universal access to ICTs has focused on stimulating the market wherever possible and avoiding subsidy (see Figure 3.2.). This applies mainly to infrastructure or mechanisms for information transfer as well as many value-added services. The market will not be fully liberalized

Table 3.1 Goals for universal access in Uganda (from UCC Rural Communications Development Policy)

Goal	Implementation mechanism(s)
Bring access to basic communications to all subcounty levels in Uganda or to every community, which has a population of at least 5,000 in habitants, by the year 2005	Licensing obligations, use of RCDF where market is insufficient, increased liberalization after end of exclusivity period
Ensure effective utilization of the Rural Communication Development Fund (RCDF) for rural communications development	Identification of underserved areas, allocation of service lots to private enterprises via least-subsidy method, subsidization of 30–50% of costs, 100% for content development
Promote ICT use in Uganda	Establish cybercafés and telecentres at district level, enable market, pursue liberalized technology-neutral policy, provide ICT training through vanguard institutions, support development of locally relevant content
Promote communications in rural areas as a profitable business	Pilot projects, subsidization of start-up of business initiatives that will be economically viable over the long term

until the exclusivity period of the two major operators – MTN and Uganda Telecommunications Ltd. (UTL) – ends in July 2005. Until that date, the licensing obligations of these operators are another tool for encouraging the growth of infrastructure and services in areas that the companies otherwise view as too unprofitable. In addition, the RCDF is being used to subsidize specific infrastructure and services in areas designated as 'unprotected' due to the stated lack of interest of the incumbent operators in doing business there. Of the three case study areas, only Lira was a direct beneficiary of the RCDP at the time of research, a 'point-of-presence' subsidized by RCDP having been established there in May 2003.

Locally relevant development-related information is a separate matter. The government sees itself as a large producer of relevant information and the overall intention is to freely distribute it. This falls largely outside the mandate of the UCC, although it is also subsidizing the creation of local content via district portals.

Regulation, licensing obligations and enforcement

The UCC, the telecom regulatory body in Uganda, drafted the RCDP and is implementing it. This policy is widely seen as innovative and progressive, a potential model for other countries (see Box 3.1).

Box 3.1 Licensing obligations for major operators

Line roll-outs
100,000–200,000 distributed at the district, then county level
  MTN: 87,000 lines distributed according to region
  UTL: 100,000 lines

Pay phones
One pay phone per operator per county
  MTN: total 2,000 pay phones
  UTL: total 3,000 pay phones

Identification of 'unprotected areas'
By July 2002, each operator had to declare which areas they were not going to serve because they are not commercially viable; these would then be opened to competitive subsidy using the RCDF.

Payment into RCDF
Each operator to pay $200,000 at commencement of licensing plus 1 per cent of gross annual revenue.

Table 3.2 Links between universal access and national development objectives (based on the RCDP and the National ICT Policy)

Development benefit	Role of ICTs
Social and economic development of country	As ICTs are vital tools, affordable basic access is critical.
Connection to the global information society	ICTs are a central component of the information society.
Effective decentralized local governance and administration	ICTs facilitate flow of information between levels of government, and between government and citizens, and increase transparency.
Improved engagement in markets	ICTs provide access to market information.
More efficient business and government practice	ICTs can reduce administrative overheads.
Sectoral service delivery in health, education, agriculture, etc.	ICTs reduce rural isolation and improves access to and delivery of these services.

Table 3.2 – *continued*

Development benefit	Role of ICTs
Multi-sectoral development	ICTs have applications relevant to health, education, agriculture, government, commerce, etc.
Economic growth	ICTs enhance competitiveness, increase trade and investment, and can help modernize the private sector through improved market access, sales, trade and knowledge of business trends.
Creation of opportunities and empowerment	ICTs provide access to local and global markets and can be used to promote rural development.

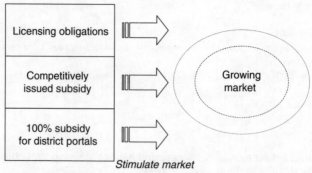

Figure 3.2 Regulatory strategies for achieving universal access

At the time of research, the market was limited because only two major operators were licensed: MTN and UTL. The companies are permitted to provide a full range of telecommunication services, including fixed line, cellular and international data gateway. This legalized duopoly is seen as a necessary incentive for the companies to invest in expensive infrastructure. Uganda was generally not viewed as a favourable place for such investment due to the high level of poverty and low levels of electrical and road infrastructure. Retrospectively, it has turned out to be very lucrative for both companies, but especially for MTN, which dominates the cellular market. As part of the licensing process, each was also required to pay US$200,000 at the beginning of the license, to pay 1 per cent of annual gross revenues, and to install basic telephone services on a geographically determined basis.

The other cellular operator is Celtel Uganda Limited (which was actually the first mobile operator, licensed in 1995). It was originally not recognized as a major operator, since it was licensed before the Act was passed and the terms had been clearly defined. Its license was later changed to recognize it as a major operator, giving it greater rights, the most important being the ability to operate its own international gateway (Tusubira *et al.*, 2003). MTN has the largest number of cellular subscribers, estimated at over 300,000 in 2003, while Celtel had about 10 per cent of that. MTN also has the broadest network coverage in Uganda. By the end of 2001, MTN was serving 85 towns covering 65 per cent of the country and 75 per cent of the population (Government of Uganda, 2002). By the end of 2004 the coverage had expanded to 120 towns and over 90 per cent of the nation's urban population.

MTN and UTL are using the exclusivity period to strengthen their hold on the market, especially with regard to the Internet services. UTL is also an Internet service provider (ISP) and there is evidence that it is using its monopoly position to squeeze out other ISPs, for example by offering a Freenet service. While UCC is 'technology neutral' in its policies, VoIP is illegal in Uganda unless one of the two major license holders is included as a third party. Likewise, no more international data gateway licenses will be issued during the exclusivity period. Afsat is one company that held such a license predating exclusivity and it is the only company that was marketing VSAT services in Uganda in 2003, through a service called iWay. An iWay VSAT terminal cost about US$4,000 and monthly service was in the range of about US$200 to US$300. This price was much cheaper than what had been available a few years earlier, when a satellite cost tens of thousands of dollars. However, this accommodated a data transmission rate of only 64 kilobytes per second and did not allow for bandwidth resale.

By 2003, 17 ISPs were licensed, with most offering services only in Kampala. However, Bushnet offered wireless services to many areas outside of Kampala, via line-of-sight transmission along the MTN GSM towers. And Afsat offered VSAT, which could be installed wherever there was electricity. Neither of these was a cheap option. The alternative was a long-distance phone call to a Kampala ISP, which provided Internet service that was both expensive and of poor quality.

Internet cafés previously required licenses but the UCC waived this requirement to encourage growth. Likewise, telecentres do not require licenses and the UCC had no formal manner of tracking them.

Non-governmental initiatives

Uganda has a high level of donor and NGO activity, since it is seen as a stable, well governed but poor country where development dollars are likely to make a positive contribution to the country's advancement. In addition, some of Africa's first telecentres were in Uganda, with donors often working in collaboration with the government and local NGOs. Four of the telecentres included in the case studies – the Nabweru telecentre, CPAR Community Learning Centre (CLC) in Lira, and the African Highlands Initiative telecentres in Kabale – were funded by IDRC. The UCC estimated there were about 14 non-governmental telecentres supported by outside donors (excluding World Links telecentres), although nobody had exact figures.

World Links is an international NGO set up by the World Bank. Its main activity has been to set up school-based telecentres (SBTs). World Links Uganda, run by former Nakaseke co-ordinator Meddie Mayanja, had established 15 SBTs in 12 districts. Most were connected via Veristar C-band VSATs, except for four in the area around Kampala that were using wireless broadband (via microwave). These telecentres were computer facilities set up in secondary schools and opened up to the public after-hours. The school is the core client, and the students contribute to the telecentre through a fee of Ush5,000 to 10,000, paid as part of the school tuition fees each term. World Links provided basic equipment and satellite connection, and subsidized the running costs for the first three years of the programme. It also provided ongoing training to the teachers running the SBT, some management and technical support, and networking and informational support including a quarterly newsletter. Schools underwent feasibility assessment before being accepted and were required to develop business plans for the SBTs.

The whole system was working well, according to the national co-ordinator, and with many schools making a surplus, sustainability was not a pressing issue. But then, in January 2003, Veristar went bankrupt. Because the equipment was proprietary, it was not easy to find another service provider and much of the original investment was lost. The schools were off-line for almost eight months while this was being resolved (they eventually went to Afsat). Fieldwork for the research reported here reported here coincided with the sixth month without connectivity and with the end of subsidization looming. While it was not the best time for a researcher to get a proper feel for the full potential of these telecentres, two of them (in Lira and Kabale) were visited during the course of the case study work.

Review of the three community case studies

Lira

'I joined the computer club because the computer is important: without it one can't make a better study [in other school subjects]. I can get extra material via Internet and get connected to friends outside'.

Student at Lango College, Lira

Lira District in northern Uganda has been subject to ongoing insecurity and civil war, which have been especially hard on rural civilians. The town of Lira has been a place of refuge and stability and has grown quickly as people flee the countryside. Traditionally, though, it has had a subsistence economy with very little local industry. It lacks adequate health facilities and food shortages are common.

From about 1998 to 2003, Lira town went from having virtually no ICT services to offering a whole gamut of them: local FM radio stations, Internet cafés, computer training, computer business services, and landline and cellular telephone services (private and public). With the exception of radio and, to a lesser extent, telephone, awareness and use of most of these services remain very limited and drop off quickly as one moves farther from the town. Except for the MTN cellular network, services do not extend into the district beyond the town because of poverty, poor infrastructure and insecurity. In town, local entrepreneurs are enthusiastic about ICTs both as a community service and as a livelihood. They have shown some creativity in marketing their services and are willing to provide high levels of user support. Many of these people are university-educated locals who have chosen to return to their home town. Their ability and desire to create a livelihood in Lira benefit the district.

In Lira town, the radio stations and public Internet centres have in some sense grown up together. Radio presenters were some of the earliest users of the Internet cafés, going there to find news and music for their broadcasts. Radio stations and cybercafés support each other by exchanging free on-air promotion for free Internet access. Radio Lira opened its own cybercafé and also became the sales point for Afsat's iWay VSAT services in northern Uganda.

Lessons
Lira town has seen immense changes in the availability and use of ICTs over the last three years. It demonstrates the potential of radio as a way to directly promote and popularize ICTs among people who might otherwise be unaware of them. This can be sufficient to spark initial interest, although in Lira it was still too early to say what the long-term effects of these promotional efforts would be.

Lira's experiences also show that, although ICT skills are sometimes seen to promote brain drain away from areas with limited employment potential, they can also provide an impetus for staying within a community. This has been the case for a number of entrepreneurs who started ICT-related businesses, focusing on popularizing their use and providing broad access. These businesses have faced challenges, especially because of limited access to capital and because of their dependence upon long-distance dial-up services for Internet access. The availability of local points-of-presence and the opening of the market to a greater range of technologies offered competitively can make a huge difference to these smaller businesses. Otherwise, those larger businesses or externally supported non-profit initiatives able to procure more expensive but satisfactory services such as VSAT (currently expensive, in part because of the single operator and illegality of resale) will be able to out-compete local entrepreneurs based simply on their access to larger amounts of capital. This limits the potential role of competition to the detriment of both the smaller entrepreneurs and their potential customers.

Lira's experiences also indicate that local businesses learn quickly from each other, can respond to their customers' needs, and are capable of providing strong support when they understand that such support is necessary for their own sustainability. In other places, there is a greater tendency to provide minimal support – to count on users already knowing the technology or being able to learn it on their own. In Lira, the low levels of capacity and awareness in the population have made this strategy less viable. Local businesses have recognized the need to grow their customer base.

Despite the outreach efforts of businesses in Lira, there are obvious limitations to the uptake of ICTs there, particularly computer-based technologies. One is the relatively high cost of services given the very low incomes of the rural majority, compounded by transportation and security issues. Many rural people rely on radio. The use of the Internet by radio presenters increases the listeners' access to international and national news and other information, and most programming is in local languages. Information on local security, relayed by reporters in the villages, is the prime concern of most listeners. And where phones can be used, these too facilitate the speedy communication of security information.

Nabweru

'Our marketing happens on a very small scale. We give introduction letters to neighbouring schools, mainly to the teachers. We haven't done more because we don't have the capacity for more customers'.

Owner, communications centre, Nansana

'Those with less education will sometimes come for computer training to get jobs because they don't have money to continue with their studies'.

Attendant, computer training centre, Nansana

Nabweru Subcounty is just north of the city of Kampala. Its largely rural population depends mainly on small businesses in trade and retail, services and production. A third of households also grow food. Depending on where they lived, people had reasonably easy access to services within Kampala and en route to it. However, there were no ICT services within the subcounty proper until 1998. At that time, a telecentre offering a full range of ICT services was established at the subcounty headquarters, funded by IDRC and managed by the UNCST and a local management committee. In 2002, the pilot phase ended. The telecentre stopped receiving funding and was turned over to subcounty authorities, which took full ownership.

The telecentre never had high usage levels; at its peak, it had about 300 users a month. Outside competition gradually increased, especially in telephone services. Dial-up Internet services were stopped in 2002 because the telecentre could not compete with cybercafés using wireless.

The telecentre never had the sort of community ownership and input that were hoped for, and it did not manage to make services widely accessible to the public. One reason for this was simply the low level of public awareness of ICTs within the subcounty. In addition, incomes and ability to pay were generally quite low, although the subcounty was prosperous relative to most of Uganda. Per-hour user fees deterred the merely curious, while computer training also was an investment too high for most to justify without a very clear notion of the benefits. Another factor has to do with the telecentre itself: its poor location. It was beside a jail and far from other shops, amenities and convenient transportation.

Meanwhile, Nabweru is home to a large number of entrepreneurially minded people. Small ICT-related businesses, especially those offering basic computer training and business services, numbered about 15 within the subcounty, not including phone bureaus. Most of these were clustered within the trading centres of Nansana and Kawempe, while similar businesses were also found just outside of Nabweru in Bwaise, including a cybercafé. Many of these businesses struggled to find adequate start-up capital, did not have a clear idea of the kind of demand they might expect for the services offered, and invested little in marketing. Many had gone out of business, although the sector as a whole was expanding.

Lessons

With the clarity of hindsight, it is easy to see that the design and implementation of the Nabweru telecentre could have been better. A more accessible location, attention to marketing and pricing, and earlier local input into substantive management issues, including expenditures, are all measures that could have helped. Still, it is not obvious that there was 'one best way' for the telecentre to address local needs. Even with the suggested improvements, the situation would have remained a major challenge, especially with regard to providing affordable services and building local demand.

It is interesting to speculate what might have been achieved had the original funds been used not to set up a telecentre per se, but to support local entrepreneurs in setting up their own ICT-related businesses. Perhaps local demand was not sufficiently high at the time for this to have been a viable option, although for phone services at least, it should have been. A loan or small-grants programme could have enhanced the viability of entrepreneurs' early efforts and increased their reach, achieving many of the same objectives as the original telecentre. The money could have been used, for example, to capitalize start-ups, train maintenance technicians, market services, network with other local entrepreneurs and share knowledge in other ways. Nevertheless, it would still have been quite a challenge for such entrepreneurs to provide truly affordable access for those on very low incomes.

Kabale

'One problem we have is that people don't know how to handle the phones, especially rural people. They have problems with timing and handling the receivers. We also have problems with people stealing phones'.

Attendant, privately owned public phone booth, Kabale

Kabale is a mountainous district in southwestern Uganda, bordering Rwanda. The research focused on Kabale town, the district's capital and business centre, and on two rural subcounties, Bukinda and Rubaya. Kabale is a densely populated district and farmers face environmental threats related to soil erosion. The area attracts a modest number of tourists because of its natural beauty and the fact it is home to endangered mountain gorillas.

The Kabale case study highlighted the vast differences in ICT provision between towns and rural areas in Uganda. While Kabale town had one cybercafé, two telecentres (neither of which had Internet access at the time of research) and numerous public phone and computer-related services, the types of services offered by these centres were absent from rural areas. The exceptions were private cellular phone ownership (but at low density) and two or three ICT projects incorporated into broader development programmes. In these areas, most people depend on subsistence agriculture and their disposable incomes are small. Moreover, roads, water supply and electrical infrastructure are generally in poor condition.

The experience in Kabale town was parallel to but also distinct from that in Lira. Both towns had early attempts to provide Internet access via long distance dial-up which proved to be of poor quality and difficult to sustain. In Kabale, a business centre at a large hotel and a subsidized telecentre had managed to provide such services, but at a loss. Both towns had school-based telecentres located a few kilometres from the town centre. Both towns had business sectors that had been fairly active in experimenting with ICT provision.

In Kabale, there were no businesses offering maintenance and repair services, and most of the business support services such as typesetting and printing had grown out of existing stationery shops. Businesses had responded to the tourism market and often communication services were located within or beside hotels. There was little emphasis on popularizing ICT services beyond the small segments of student and professional groups who already knew of and used such services. The exception to this was the African Highlands Initiative (AHI) telecentre, since it had a mandate to serve rural populations and contribute to local development. However, its staff did not have any clear strategies for attracting the general population, had at that time no resources for a marketing or advertising campaign, and had found that few local farmers and non-professionals made use of their services. Money, distance, interest and awareness were all widely noted as important factors that restricted use of the telecentre.

Bukinda and Rubaya are rural subcounties that depend heavily on subsistence farming. Each hosted a rural ICT project. The World Vision office in Bukinda hosted RANET, a project in partnership with the Ugandan Department of Meteorology and the international RANET consortium, which delivers climatic and other information via a satellite. Rubaya had its own AHI telecentre located beside the subcounty headquarters, linked to the telecentre in Kabale town. Both of these had experienced technical problems that limited their service offerings. In Rubuya, for example, the telecentre could not establish a wireless Internet link as originally intended and the cellular phone network worked poorly. Insufficient local technical capacity also meant the telecentres had to depend on relatively inaccessible external support which tended to be slow in coming. Nonetheless, as far as these centres had been able to integrate with other development activities, they had provided interesting extra capacity. In Rubaya, such capacity was due more to the organizational skills of the staff than to any ICT use. In Bukinda, the RANET site allowed for the reception of climate forecasts and other information of use to local farmers.

Lessons

Arguably, the most important lessons from this case study relate to rural awareness and use of ICTs. While the ability to communicate and interact with people in other locations is often highly valued in isolated rural areas, many factors inhibit use of ICTs. Information-seeking behaviour does not occur spontaneously if people are not already accustomed to the availability of information and if they have their own methods of decision-making based on custom and habit. Creating an access centre, especially a multi-purpose centre, in a rural area thus implies much more than simply overcoming the great obstacles to building and maintaining an infrastructure in an environment where supports are few and far between. It also implies creating new organizational and social structures, and encouraging new behavioural patterns that allow rural residents to seek out and gainfully apply information. Starting such a venture from scratch is a complex, long-term proposition that may also be expensive and risky. Predicting the venture's specific gains to the local community can be rather difficult.[1]

The experiences in both the Bukinda-RANET and Rubaya telecentres support the argument that a more organic, gradual approach is needed, namely one that builds on what is already there (in terms of both organization and physical infrastructure) and uses ICTs to augment it. The experience of both telecentres also supports the argument that user-pay systems are completely inappropriate for rural ICT projects, at least in the early stages. This is because rural people cannot afford to pay – and cannot be expected to pay for what they do not yet value. This is especially true for information-related services, as opposed to human communication services, like the telephone, which rural people tend to value highly.

RANET has focused on a specific informational need and has trained farmers how to apply the information. When the system is in place, farmers notice tangible benefits to themselves within one season. They are better able to plan for and cope with climatic conditions, resulting in more plentiful harvests.[2] But an additional, higher-level benefit is that they have learned how to obtain, value and apply practical information delivered via ICTs. This is a skill, a behavioural change, that can help them improve many other aspects of their livelihoods.

Comparing the Ugandan cases with access-related development objectives

Table 3.3 gathers observational evidence from the three case communities and relates it to the eight development benefits that Ugandan policies indicate ICTs can provide. Overall, it shows mixed movement towards the intended benefits. In many instances, these observations are tentative without further economic data – for example, on the contribution of ICTs to local business.

Table 3.3 Links between the intended benefits of access policy and conditions in case study sites

Intended development benefit	Situation observed in case study communities
Connection to the global information society	All three case studies show similar patterns: ** The vast majority of people have never heard of the Internet. ** A small minority of young, educated adults use the Internet primarily for e-mail and, to a lesser extent, to gain access to international information, especially related to their studies and NGO work. ** SBTs allow students to participate in international fora and online exchanges, limited by equipment and staff availability.
Effective decentralized local governance and administration	*Nabweru* There is some degree of support to the subcounty headquarters, mainly through computer typesetting and telephone communication. *Kabale* Rubaya subcounty headquarters were located near the telecentre; district staff had access in Kabale town. While actual use remains low, some civil servants reportedly do use the services in their work (mainly administrative tasks). *Lira* Local district civil servants had access to ICT services and some were using them. The Ministry of Local Government was in the process of setting up a database system for local government and training staff in its use.
Improved engagement in markets	*Nabweru* Mobile phones and public phones are used by business people and can facilitate this. However, there is not sufficient evidence to draw clear conclusions. *Kabale* Within and around Kabale town, telephones have improved contacts within local and national markets. Benefits are more limited in remote areas, although the potential is there. RANET participants in Bukinda hope to link the climate information to marketing strategies.

Intended development benefit	Situation observed in case study communities
	Lira Use of ICTs for this purpose was very limited in the agricultural sector, although sometimes phones were used for this purpose.
More efficient business and government practice	*Nabweru* This has been achieved to some degree through the provision of phone service. Computers are used largely as advanced typewriters.
	Kabale Phones and photocopying are broadly used, where available, to support more efficient practices (communicating, recording and distributing information) while computer-based ICTs and fax remain limited and mainly used for word processing activities.
	Lira Some civil servants were using ICTs, although this was still fairly novel. Staff of public access centres characterized NGOs and civil servants as the main users, businesses less frequently.
Sectoral service delivery in health, education, agriculture, etc.	*Nabweru* There has been very limited use of ICTs for this purpose. Some professionals make use of the Internet at cafés, including teachers and health professionals. Agricultural information delivery programmes at the telecentre have been of limited reach and duration.
	Kabale Public access centres support individual professionals within these fields who choose to use the services. The SBT at Kigezi High School offers some students enhanced educational opportunities (in international exchanges and projects). ICTs have improved delivery of international news services. Agricultural uses have not flourished, with the exception of RANET, delivering climate information.
	Lira Civil servants from these sectors had better communication with Kampala offices and more streamlined administration. Phones could be used for health emergencies. The SBT at Lango College offered some educational opportunities to students (but limited because of equipment and staff shortages).
Multi-sectoral development	*Nabweru* The impact of ICTs on local enterprises and livelihoods is not clear from the data gathered, although telephones have been adopted by many small local businesses, especially in trade-related areas.

Table 3.3 – *continued*

Intended development benefit	Situation observed in case study communities

Kabale

Uptake of ICTs has been limited in most sectors, especially small businesses and farming; so its impact on the development of these sectors is likewise limited.

Lira

The sector most affected appeared to be the news media. Both radio and newspapers were benefiting from increased information flows.

In all three:

Economic growth

- Proliferation of ICT-related businesses and business expansion suggest that ICTs have contributed to local economic activity. The contribution of ICTs to local wealth generation is less certain. While phone services are usually viable, computer-related services have low profits and many businesses do not survive. Further, the equipment and many of the operating services (ie, connection, repairs, consumables and electricity) are bought from external suppliers. Economic activity directly due to ICTs in town is large, but because money leaves the community, economic growth may be marginal. ∶
- Access centres provide some local jobs.
- In more remote towns, the opportunity to begin local ICT businesses may help to retain skilled and educated local entrepreneurs.
- If ICTs help retain the elite who tend to use these services (it is too early to tell if this is the case), then they may contribute to local economic growth because these are people who can afford to invest locally.
- Telephone, fax and Internet allowed local business people to communicate and do business with those in other centres, especially Kampala, facilitated where the road system and security were good.

Creation of opportunities and empowerment

- Where there are increased local services, most of the population lacks the awareness and capacity to profit from them. Those who do are overwhelmingly wealthier and more educated, thus evidencing localized 'digital divides.'
- For those who live in towns and can pay, ICTs have offered a range of opportunities – for example, for Ugandans based in Kabale engaging and monitoring in international investment activities without ICTs.
- In rural areas, ICTs offer opportunities only where they are integrated into broader development programmes, such as RANET.

Intended development benefit	Situation observed in case study communities
	• The growth of ICT as a sector and the provision of ICT services allowed those from Lira but educated elsewhere to seek local economic opportunities and bring their education back to their hometown. This opportunity to stay home was important to the whole community, as these were people who had much to contribute both economically and socially.

Another important overall observation is that this situation was clearly dynamic. There was a clear divide between towns and villages in the level of awareness and use of telephones, but many village families had relatives working in town. These people would often introduce their rural relatives to telephones or even buy them cellular phones, and this trend meant the divide might be transitory. With computers and the Internet, this was much less likely, since these required a reliable source of electricity, maintenance, and literacy to use.

Learning from Uganda's experience

Liberalization, cost and technological availability

Uganda's decision to introduce a second national operator before completing privatization of its state-owned operator has meant that it has had more competition in its telecom market than most other African countries. Under these conditions, fixed-line service has improved greatly although cellular penetration has increased much more quickly. The spread of fax and the Internet are both limited to fixed-line penetration. Wireless broadband use has been very popular as ISPs can offer it directly. However, it can only be practically used within a relatively small distance of an international data gateway. Its use, therefore, is widespread in Kampala but rare in other towns.[3]

Opening the market has created relatively more opportunities than in South Africa. However, many technologies, such as VoIP, are still effectively unavailable or only available through a limited number of operators, under certain conditions. And prices remain high. Cellular phones and public phones are affordable in the sense that a single service unit can be purchased at a time (ie, pay-as-you-go). Nonetheless, the tariff rates remain high, and for the rural majority, phone isn't affordable.

The expense of connecting to the Internet through international gateways limits bandwidth availability and keeps costs high. Some people are working around this by maximizing the within-country connectivity, which is potentially much cheaper. Bushnet, for example, has a system that manages traffic to maximize the within-country speed and bandwidth and reduce international bottlenecks. Others have spoken of creating within-country broadband networks that minimize their use of the international link.

In Uganda, radio and telecommunications were liberalized during the same period. This has had some interesting outcomes, especially in towns far from Kampala, where there has often been a convergence of ownership and interest between the two. Specifically, some radio station owners have introduced on-site cybercafés, promoting them through their broadcasts.

Reach of centres and the urban-rural divide

In Uganda, the decentralized government structure, including the wide spread of district capitals, encourages a broad but superficial distribution of ICTs across the country. This has been further encouraged by the relatively high level of liberalization and by explicit UCC policies that set reach targets by district and subcounty. Despite this spread, there is a big gap between towns and even the closest surrounding villages in terms of awareness and use. This gap has not been bridged, although in some places, such as Lira, people are trying. School-based telecentres, often located on town outskirts, also increase accessibility for the neighbouring populations.

Providing access in rural areas is a big challenge. On the supply side, limiting factors include infrastructure, maintenance, equipment supplies, staff capacity and transport. On the demand side, several things are lacking: money, awareness, client transportation to the centres, and clarity (sometimes on the part of the centre itself) about how the services can be used to improve livelihoods.

The experiences of access centres in this study suggest that user fees are major barriers in popularizing services. Within rural areas, one of the best ways to overcome inhibiting factors is to embed access to and use of ICTs into existing development processes. RANET is an example of this. It helps to absorb the costs and complications of supply and to address demand limitations, by using existing programme delivery mechanisms, relationships and organizational structures. In developing access centres, especially in rural areas, small steps are best. The more remote and difficult the situation, the more important it is to build on what exists and to break down and deal with complexity in stages.

Strategies for the short, medium and long term

Strategies to increase access to ICTs must run along a number of different timelines and be harmonized with other efforts. Infrastructure development, especially fixed line and electricity, are important factors that can limit the effective supply of telecommunications. Lack of awareness and capacity amongst the general population limits both supply (ie, people able to develop, offer and maintain services) and demand (people aware of and able to use services). The UCC is already using its funds to sponsor development of vanguard ICT-training institutions within different districts as a way to build this capacity. Over the long term, equipping schools, training teachers, and developing appropriate curricula are important strategies that are being pursued by both the government and NGOs (such as Schoolnet Uganda).

Within the different timelines, the specific role of shared access centres, and specifically those that are subsidized, is dynamic and requires some careful thought. Where centres are set up under the most adverse conditions, they often fail because they do have the vast resources needed to confront those conditions. But attempting to surmount the obstacles may not actually be a top priority for the communities in which the centres are located. So we have a conundrum: since community members are unfamiliar with ICTs and their potential uses, they are unable to make an informed decision as to how valuable a telecentre-type service would be. Arguably, at least some of the inhibiting factors, such as lack of electricity, should be first addressed by separate efforts so that the complexity and expense of introducing ICT services is reduced to some manageable level. At the other end of the spectrum, some development-funded efforts have established public access centres in places where the market was already effective. Or they have subsidized or donated expensive equipment to local organizations, including businesses. Such anti-competitive behaviour is unfair and as likely to inhibit local development as to support it.

This dilemma has no magic solution. But it does stress the need for reflection before the establishment of funded centres. Feasibility has to be balanced with need, and in practice the two often have a strong inverse relationship. The purpose, target users, differentiation from other services, and the longer-term vision for the centre all require forethought, with the understanding that the ICT sector is dynamic and that some 'rural' areas, although far from Kampala, nonetheless host entrepreneurial centres.

Developmental impact and potential

That the developmental impact of ICTs depends entirely upon how they are used is almost a truism. Yet it bears repeating. Use in turn depends on the needs, capacity, inclination and awareness of the user, plus the larger organizational and social structures in which the ICT is embedded. The Internet can be used to find pornography or to find information useful to basic food production. The telephone can be used to make a life-saving call during an emergency, to contact Short Message Service (SMS, which allows messaging to mobile phones and other wireless devices) for current market prices on food commodities in Kampala, or as an expensive social status symbol. One use does not preclude the other; indeed, even if a phone was purchased to make a social statement, it still opens the potential for uses with tangible value.[4]

In Uganda, trends in Internet use favour it as a tool for sending e-mail, especially internationally. The World Wide Web is most often used as a source of entertainment, music, pornography, education, information on overseas educational opportunities and scholarships, job opportunities and international news. Some professionals also use it in their work-related activities.

F.F. Tusubira, a commissioner of the UCC, has noted that cellular companies found the peak use of phones was between 4 pm and 9 pm, strongly implying social use. This further implies that most phone use does not create economic value for the user, but rather produces profit for the large companies at the expense of poor Ugandans. This research also found that social use of the telephone, especially to contact family members, is the primary use. However, because economic interdependence amongst family members is such an important part of most people's livelihoods, such use is not necessarily non-economic. It is very difficult, however, to judge the economic value of telephone use versus cost – there is no clear linear relationship. It is also apparent, in both South Africa and Uganda, that social factors, as much as or more than economic factors, motivate people, particularly young people, to acquire a cellular phone. It is very rare, for example, to see a Ugandan owning the cheapest model of phone available in the shops; indeed, many people own higher-end models.

Public access centres remain the main vehicle for ICT use for most people in Uganda. They can also be important sites of socialization around ICT use. Pure public use (where the centre sells a service) tends to mirror the patterns people have learned in using ICTs at university or in work settings, which often include entertainment-related uses. Purpose-driven ICT projects, such as RANET or specific online collaborations amongst students at an SBT, are rarer and generally target users who have not been socialized to ICT use in other settings. They also tend to embed the use of ICT in activities that have a specific, intended value.

CHAPTER FOUR

Start-up and scalability of access centres

This chapter examines the related issues of access centre scalability and sustainability. Scalability can be considered at a micro-level as the question of how a centre actually starts up. From a national policy perspective, this means looking at what favours a successful start-up and how best to create these conditions. Sustainability is the issue of how a centre manages to continue operating, whether through user fees or some other means, and the types of conditions that may put a centre at risk of closing or compromise its services. Again, from the perspective of national policy, the main issue is how best to support conditions in which access centres have a reasonable chance of surviving and operating in a way that their services will be useful and accessible.

In the course of this study, about 65 different access centres were encountered. While most were within the case study communities, about 10 more were from different areas of South Africa and Uganda. The extras were included because they in some way expanded the lessons and experiences from the case communities.

Access centres can be broadly placed in two categories: for-profit and not-for-profit. The former grouping comprises phone shops, Internet cafés, communication centres (including computer training establishments), and services within an existing business such as a hair salon, traditional healer or stationery shop. Not-for-profit public- and NGO-sector services include those within an existing not-for-profit organization, those that are part of an integrated development strategy, and telecentres. The access-related services offered by these centres are here broken down into five types:

1. direct access to one or more ICT tools, often including basic customer support;
2. training and skills building in the use of computer software packages, sometimes for business management;
3. access to climatic information, market prices and other information;
4. information production or processing such as desktop publishing, video editing and graphic design;

5. some other kind of activity augmented by ICTs (access as a means to an end).

Table 4.1 correlates the seven types of access centre with the five types of access-related services.

The rest of this chapter explores how the different types of access centres began and sustained (or failed to sustain) themselves, and why they offered the types of services they did. It also compares and contrasts communities with regards to the types of access centres available (see Table 4.2).

For-profit sector

Private phone shops

'I'm the owner and I got the start-up money from selling land. To me, the greatest success of this centre is my marriage – I got my wife through this centre'.

Owner/operator, public phone booth, Kabale

Privately owned and run phone shops existed in all case study sites except Bhamshela and rural areas of Kabale. These shops often consisted of nothing more than a single metered phone on a plastic table set up outdoors. Photo 4.1 shows a wooden booth on the side of a street, housing a metered wireless phone and an attendant with an account book and a customer. This was a

Table 4.1 Services offered, by type of access centre

	Direct access	Skills building	Access to specific information service	Information production or processing	Other activity augmented by ICTs
Phone shops	often	rarely	rarely	rarely	rarely
Communication centres	sometimes	often	rarely	often	rarely
Internet cafés	sometimes	sometimes	rarely	often	rarely
Services within a business	sometimes	rarely	rarely	often	sometimes
Services within . . .					
School	often	often	rarely	rarely	often
Library	often	rarely	rarely	rarely	often
NGO	sometimes	rarely	sometimes	rarely	often
Community centre	often	often	rarely	rarely	sometimes
Telecentre	often	sometimes	rarely	often	sometimes
Integrated into development activity	sometimes	sometimes	often	often	often

Key

☐ rarely   ▨ sometimes   ■ often

Photo 4.1 Privately owned phone shop on Main Street, Lira

typical set-up in Lira. In some instances the phone shop was integrated into an existing business, such as a hair salon. Almost all of these used fixed wireless technology rather than landlines. There were often many of them in a single town, generally clustered around high-traffic areas such as taxi ranks and shopping centres. Typically, they advertised their service and rates to passers-by via simple outdoor signboards.

Phone shops were popular with entrepreneurs, most of whom lived in or close to the area where their shop was located. These enterprises cost little to start and run. Owners usually hired attendants and sometimes opened phone shops in various locations. While attendants often worked long hours for low wages, the shops were at least able to cover the costs of both the staff and phone line and make a profit. Owners were almost always motivated merely by profit. In interviews, some said they had considered expanding into other ICT-related services, but this would normally require a great deal of invest-ment, especially for those who were simply operating on the street. As most were reasonably satisfied with their profits, they had little immediate motive to expand.

While most phone shops offered only basic phone service, a few also provided connection time and phone cards. Often the attendant would dial the number for the customer. Because charges were typically 'per unit' and the attendant often did not know how long a unit lasted (it was tracked automatically on the metering system), the charges and charging method were generally not transparent to customers.

Table 4.2 Types of services available in five case communities

Case	Centre type	For profit?	Number	Year started	Sample services
				South Africa	
Bhamshela	Universal Service Agency telecentre	no	1	1998	phone, fax, computer access, computer training, Internet
	MPCC	no	1	2003	government services: birth, death, marriage registration; social security cheques; employment assistance; educational information
	services within other organization (Village Bank)	yes	1	2002	fax, photocopying, possibly Internet
	phone shop franchises	yes	2	2001	phone, phone cards
Khayelitsha	Universal Service Agency telecentre	no	1	2000	photocopy, fax, computer training, computer access, typesetting, printing
	phone shop franchises	yes	many	1998 onwards	phone, air time cards
	services within existing businesses	yes	2 or 3	2000 onwards	photocopy, fax, typesetting
	services within existing NGOs	no	1	1999	photocopy, fax, typesetting, e-mail, computer access, printing
	private computer training centres	yes	2	?	computer training, mainly on commercial software packages
	NGO computer training	no	1	?	computer training and entrepreneurial skills-building, aimed at starting own desktop publishing businesses, etc.
	services within library	no	1	2000	Internet (4 other libraries with photocopying)
	independent phone shops	yes	many	~2001 onwards	calls to all networks
				Uganda	
Lira	private communication centres	yes	6	1997 to 2001	computer training, computer repair, consultancy, photocopying, typesetting, binding, printing, stationery sales, fax, e-mail
	independent phone shops	yes	many	~2002 onwards	local, regional and international phone services, phone cards
	private Internet cafés	yes	2	2003	Internet with staff assistance

Location	Centre	Electricity	Number	Started	Services
	World Links telecentre	no	1	2001	computer training, computer and Internet access
	CPAR Community Learning Centre	no	1	2003	planned services to include Internet, video and audio editing, secretarial and desktop publishing services
	post office	yes	1	about 2000	fax, phone
Kabale town	private communication centres	yes	12	~1999 onwards	desktop publishing, typesetting, video rental, tape and video editing, scanning, video games, phone, graphic design, compact disk (CD) burning
	private Internet cafés	yes	1	2003	Internet, printing, typesetting
	private computer training centres	yes	2 or 3	2002 onwards	basic software packages
	Christian Ministry Vocational College computer training	no	1	1998	computer classes (mainly directed towards secretarial skills)
	private phone shops	yes	20	1999 onwards	phone (fixed wireless)
	World Links telecentre	no	1	2000	public computer and Internet access
	AHI telecentre	no	1	2000	photocopying, typesetting, computer training, phone, agricultural content production, links/support to rural telecentre (in Rubaya)
Kabale rural	AHI telecentre (Rubaya)	no	1	2000	typesetting, printing, meeting space, projector, photocopying, phone, computer training, small resource library
	RANET site (World Vision area office, Bukinda)	no	1	2001	agricultural information, especially climatic information, disseminated through farmer groups
Nabweru	IDRC/UNCST telecentre	no	1	1999	computer training, typesetting, computer access, agricultural information and outreach, photocopy, phone, fax (not working), TV, video library
	private communication/training centres	yes	8 or 9	1999 onwards	phone, fax, photocopy, typesetting, computer training (software packages), desktop publishing, binding
	private Internet cafés	yes	1[1]	2002	Internet, printing, typesetting
	private phone shops	yes	many	~2002	phone: national, East Africa, sometimes international

[1] This café was just outside the border of the subcounty.

Since most phone shops were very basic, there were few service features on which to compete. Location was probably the biggest single advantage one shop might have over another, although privacy, reliable and honest billing, and good rates were also factors customers tended to consider. Because the phone service was purchased by entrepreneurs from a major telecom operator (usually on a prepaid basis), there was little room for competition on pricing. This motivated some owners to surreptitiously shorten the length of a unit – so that it looked like prices were cheaper when they were not. The ease of entry into this kind of service provision meant that competition was high and, despite high demand, profit margins low.

The start-up of these types of phone shops often lagged slightly behind other types of phone access that had been offered through either the licensing obligations of major operators or public centres and larger communication centres. This is probably because the entrepreneurs running these businesses had relatively little capital and little tolerance for risk. The biggest barrier to start-up was the absence of a phone network and electricity.

Phone shop franchises

'I want to get phones in but I have to wait a month or two for MTN, and [I] don't want to. But Vodacom is full in Site C, Cell C's network is bad, Telkom is too expensive and most people these days want to call cell phones, so I'm not interested in them.'

Entrepreneur on opening up a public phone shop in Khayelitsha

Phone shop or phone booth franchises, present in both South Africa and Uganda, were due almost entirely to the licensing obligations of major operators. Vodacom, South Africa's largest cellular service provider, had created its own model of phone shop franchise that turned out to be highly successful. Although it initially required subsidization by Vodacom, by 2003 the programme was self-sustaining and generating significant publicity for the company and a potential customer base in the communities where it was operating.

A Vodacom phone shop was typically housed in a shipping container painted green with a large sign outside. It contained between four and eight fixed wireless phones attached to a central meter operated by an attendant. Entrepreneurs could buy a container for a small sum, which included a set amount of prepaid connection time to the Vodacom network, which they would replenish as they needed. Vodacom also gave the initial group of entrepreneurs training and access to credit for shop start-up. All Vodacom containers encountered in this research (one in Bhamshela and several in Khayelitsha) were well used and generated good profits for their owners.

Arguably, one of the big successes of the Vodacom container model is that it did not require much planning on the part of the local entrepreneurs: the system of pre-paying connection time (which is then resold to clients) means that one cannot get into trouble running up debts. Vodacom subsidized the rate on the connection time sold to the owner, meaning that she or he had a decent profit margin. No other cellular operator gave preferential rates to their franchisees, meaning that those working with other networks or independently were making much smaller profits. Telkom, the major telecom operator in South Africa, also franchised phone shops but the profit margins on these were too small for entrepreneurs to make a living. The owner of the Telkom container in Bhamshela explained that he had opened it only because he saw a need for such a service in his community. He made his income through other means, and generated virtually no profit from the container.

Internet cafés

'We do get some illiterate people from the country and we try to help them. Some people panic with the machine'.

Attendant, Internet café, Bwaise

While Internet cafés certainly exist in South Africa, they were not present in either of the case communities, and typically existed only in big cities such as Johannesburg, Cape Town and Durban. These cafés catered mainly to middle-class people, and sometimes university students. Prices were generally high, up to R1 a minute. Telkom had begun offering ADSL in selected areas, usually big cities, and this brought the cost of Internet access down. In Khayelitsha and Bhamshela, the only real option was dial-up, but this had proven too expensive for entrepreneurs to offer and recoup their costs.

In Uganda, Internet cafés existed in cities and mid-sized towns. Most of the cafés in towns were newer and benefited from more recently available connectivity options. These connections included VSAT, wireless broadband and local PoPs that had been set up by UTL, subsidized by the Universal Service Fund through the Rural Communication and Development Policy (RCDP). Earlier attempts to create Internet cafés (or other types of centres offering Internet access) through long-distance dial-up had proven to be unsustainable since the cost was very high and service poor.

Unlike phone shops, Internet cafés took a fair amount of capital to start up, since they required a reasonably large space plus computers and the Internet connection. They also required a higher level of technical expertise to set up and run than a phone shop. They were often begun by individuals who already had a successful business and could use the capital from this to finance them in part. In the towns especially, it was common for FM radio stations to open Internet cafés. This worked well because the stations were usually profitable and able to finance and advertise the cafés over the airwaves. Moreover, presenters could use the Internet to get information and music for their shows.

Because most Internet cafés were new and had better, cheaper Internet connectivity than their predecessors, it was not clear how sustainable they were. High overheads and narrow profit margins made the business fairly risky. In bigger towns and cities, the demand was evident but competition was also high. Isaac Kasana, who had been associated with an ISP, noted that most of their client Internet cafés, based in Kampala, had gone bankrupt or ceased operation. Despite the volatility, many new cafés were opening. As with the phone shops, a major challenge for such cafés was that, while they had many competitors, they shared many of the same fixed costs, which kept profit margins very low. In Mbale, for example, the price war between the five or so Internet cafés was so fierce that most were offering rates below the cost of provision. One of the business owners said he was trying to organize with the others and discuss setting the prices so that they could survive. In the communities reviewed in this research, most Internet cafés focused mainly on Internet services with some level of client support. While other services such as typesetting were sometimes available, they were not much emphasized by owners or requested by clients.

In Lira, in northern Uganda, the Internet cafés were actively marketing themselves over the radio and on large signs. In this region, overall awareness of, and experience with, the Internet was low. Even relatively educated people often needed a high level of support from the staff. One Internet café, opened by Radio Lira, used the airwaves to encourage people to come in and get e-mail addresses, emphasizing that no previous experience with computers or the Internet was necessary. It reported that about 20 people a day, some from surrounding rural areas, were visiting just to see the place and make inquiries in response to the radio ads. However, at this and other Internet cafés in town, user levels were modest. In contrast, the cafés in Kabale town and in Bwaise (on the border of Nabweru subcounty, on the periphery of Kampala) were well used and sometimes had line-ups. Their owners put less emphasis on building awareness, probably because word-of-mouth was sufficient.

Communication centres (some providing computer training)

'As service providers, we have a problem because the [dial-up Internet] connection is so slow, the line drops, the client doesn't get service, and you end up absorbing the loss. Most people who started the service had been in Kampala and wanted to bring the service home [to Lira]. They got high bills, were disconnected, and were out of business'.

Communication centre manager discussing the problems of
dial-up Internet service, Lira

'In Khayelitsha, individuals get computers to start businesses producing CVs, but it isn't cheap and people fade away. Customers don't have money but they have high expectations'.

Owner of communication centre/phone shop, Khayelitsha

Communication centres were the 'jacks of all trades' for ICTs, the market-based equivalent of telecentres. They were usually set up by local entrepreneurs who had some ICT training, usually in computers or business, and who perceived local demand or need for ICT services. They were often inspired by similar centres they had seen elsewhere. The services these centres offered often evolved based on the skills and equipment the owner managed to acquire and the market response. For example, if phone service proved too costly due to light local demand, the centre might cut it and introduce binding and lamination instead as a way to survive. If there were relatively few people interested in computer classes, the centre might shift to consultancy services.

In the case communities reviewed in South Africa, there were no such centres – with one exception. In Khayelitsha, a man named Tony Lamati, who owned a Vodacom phone shop, expanded his services to include fax, photocopy, typing and the design of funeral programmes and CVs. Other local businesses that he knew had tried to offer similar services had not survived. In his case, phone service profits were keeping the business healthy, while the other services did not generate much profit and may even have been running at a loss. The only other businesses in Bhamshela and Khayelitsha that offered ICT services other than phone did so as an addition to their core activities. The reason for this was that demand was light and overhead was high for other services. As Mr. Lamati remarked, people had high expectations but were not willing to pay much. As soon as a particular service was known to be in demand, other entrepreneurs mimicked it. For example, at least three places in Khayelitsha specifically advertised rates for producing funeral programmes and CVs.

Communication centres did exist in other communities, but they were high-risk ventures. Getting the mix of services right could be difficult, and any errors of capital investment in required equipment could be costly. In South Africa, the model of ICT services within an existing business appeared to be much more likely to succeed. The scarcity of businesses offering Internet-based services could not be attributed entirely to high telecom costs since there were also few centres offering other forms of ICT services such as desktop publishing, especially compared with the Ugandan case studies.

In Uganda, there were many more stand-alone communication centres in the communities reviewed, especially in towns and trading centres. This seemed due, at least in part, to a generally high level of entrepreneurship. Opportunities in the formal sector were few and wages generally low, even for people with post-secondary education. For this reason, many people saw more opportunity in creating their own businesses. Those people with computer and business skills often saw provision of ICT access and related services as an excellent opportunity. The sector was just opening up, many places were still underserved, and they could put their skills to use. By 2003, there were enough examples of existing businesses around to provide inspiration to others. Most people who had gone to university had done so in Kampala, and some were struck by the presence of Internet cafés and communication centres in the city that were not yet available in their home towns.

In contrast to people who began phone businesses, many of those who began communication centres claimed to be motivated not only by profit potential but also by the idea that the services offered would help their communities. Many had ambitious visions of centres that would offer a range of information and communication services, often focused on the Internet. But the hurdles were many, either related to business start-up or to the fact that dial-up, the only Internet connectivity option available, was too expensive to maintain.

Across the different communities in Uganda, one or a small number of people commonly acted as 'pioneers' by taking the initial risk. These people tended to have higher skill levels and a clear vision of what they hoped to achieve. The centres they started usually sparked interest amongst other entrepreneurs, with a flurry of similar centres opening over a short period. Some 'copycat' centres were poorly planned and run and chronically underfunded. Many failed. Because these centres opened so quickly after the first one, they created fierce competition in a market which was often quite undeveloped.

In Nabweru, for example, about 15 communication centres were clustered in trading areas in and bordering the subcounty. These began opening from about 2000 onwards. Six were reviewed in this research. Of these, two were reportedly making a modest profit, two were breaking even, and the other two were losing money. Although a number of communication centres had attempted to offer computer training, they did not attract many students, and competition kept prices low. Many businesses reported plans to expand their services. However, such intentions were not based on market research, since some competitors who already offered those other services noted a lack of customers.

These centres needed to be quite flexible as the market shifted. In Lira, for example, although early centres had done well by offering phone services before these were widely available, the situation changed once the MTN network reached the area. These same centres then had to change their service mix, reducing their reliance on phone and bringing in new services. During such transition periods, some centres failed, while others reinvented themselves and survived. Their fate depended a lot on the centre's capacity and on good, flexible management. Most centres had a simple organizational structure and were managed by one person or a partnership.

A big challenge faced by communication centres (and Internet cafés) was obtaining capital for investment. Bank loans were not popular as interest rates were generally high and many people would not in any case have met the conditions for loan approval. Rather, people depended on their own savings or soft loans from family or friends. These tended to be small amounts of money. While some businesses were able to expand over time, in other cases, lack of funding seemed to chronically limit the growth of the business and weaken its overall sustainability. This was particularly visible with connectivity options: those with access to sizable amounts of capital could install a VSAT, a much better long-term option for Internet connectivity than long-distance dial-up but with high initial costs. Plenty of smaller businesses aspired to offer Internet service; but, unless there was a local Internet PoP, they were unable to afford it.

Services within an existing business

'I saw a need for services, especially for kids at schools nearby who need to do their assignments. So we started with photocopying, then CVs, then computer programming and fax services'.

Xolandi Hatana, manager of New Era Driving School

Hybrids between communication centres and other types of business were common across all communities included in the research. These were usually set up by the owners of existing businesses who perceived a demand, or potential demand, for ICT-related services. Sometimes they were natural extensions of existing services. For example, many stationery shops in Kabale town had expanded into photocopying and typesetting, and sometimes into desktop publishing and binding. Sometimes, they were odd combinations – for example, a driving school in Khayelitsha that offered photocopying and desktop publishing; traditional healers in Randburg with fax services, and hair salons in Nabweru with phone services. Some owners said they liked to offer these extra services because they attracted extra customers who might then make use of their mainstay services. This was especially important where the profits realized on ICT services were slim.

Sometimes the businesses were able to use the ICTs for their own internal needs as well. Since the businesses were already established and required little extra space or staff to run the services, they needed only to invest in the equipment. Some owners found that the ICT services generated only a modest income, but once they had the equipment, there was little reason to stop if operating costs were low. Also, the existing business provided a buffer that allowed them to take a loss while they tried to build up a customer base. Some said they were happy to provide a service of value to the community even if they did not realize a return on it. In this sense, these services were quite robust and sustainable, especially in comparison to stand-alone ventures such as communication centres and Internet cafés. Some services were rare, especially Internet access, since the operating costs were often much greater than the revenues they generated. Businesses that tried offering such access usually ended up halting the service and using the computers for other purposes such as typesetting.

Not-for-profit sector

The not-for-profit ICT access services were generally set up by a national or international programme, or through partnerships, rather than as purely local initiatives. These partnerships often created unique opportunities for the centres to offer specific services or to set up in places lacking basic infra-structure. But they also posed risks, especially around the issue of decision making. Many decisions were made by national rather than local bodies, or sometimes by international bodies. In some cases, the decisions were slow to be made and inappropriate for the local area. This problem was reported, for example, in relation to equipment procurement and attempts to resolve tech-nical maintenance issues. At the local level, lack of capacity and the political manoeuvring of board members motivated by self-interest rather than com-munity benefit were frequently cited problems. Another major stumbling block for these centres was the optimistic assumptions that local demand for ICT services would be relatively easy to develop and that services could be largely or entirely sustained through user fees. These assumptions were rarely true.

Hard-won lessons from early initiatives, however, were helping to inform more recent efforts. In Uganda, many people who had been involved in the early telecentres were critical of various aspects of the experience: the use of expensive outside consultants; delays in receiving funds and setting up infra-structure; political problems within local boards and committees; and poorly defined sustainability plans. Nonetheless, many of these people remained committed to the general idea and had gone on to establish new telecentres. For example, Meddie Mayanja, the original national co-ordinator of the Nakaseke telecentre – Uganda's first and perhaps most famous telecentre – went on to become the World Links for Development Uganda co-ordinator, putting into practice many of the early lessons. In South Africa, the Universal Service Agency had stopped implementing telecentres after 2000, in part because of the numerous problems encountered by existing ones. In 2003, they were in the process of assessing the situation and deciding how to pro-ceed. Greater emphasis was being placed on feasibility assessments, but it was not clear how much institutional learning had actually taken place since many staff, especially managers, were fairly new. Their plans remained ambitious, key definitions were still not in place, and the high cost of landline phone connectivity remained a daunting barrier.

Services within an existing not-for-profit organization

Offering public or semi-public ICT access through an existing not-for-profit organization is a common approach with many different faces, shaped by the characteristics, especially the structure and goals, of the host organization. Examples of this type of set-up that were encountered in the course of the research include:

- World Links Uganda school-based telecentres (in Lira and Kabale, Uganda);
- South African schools with plans to provide public access;
- libraries with Internet access (Khayelitsha, South Africa);

- community centre with public ICT access (Aleksan Kopano Resource Centre in Alexandra township, South Africa);
- NGOs with public ICT access (Khayelitsha Educational Resource and Information Centre (KERIC), South Africa);
- the RANET Uganda programme, which worked from World Vision field offices.

There are considerable differences in the ways these organizations began offering ICT access. A brief sketch of the main issues arising from the establishment and maintenance of each is presented below.

ICT access in schools

'The problem [of break-ins] in the school environment has not been as bad as in the telecentres. In schools, the problem is with the cost of the Internet connection. For example, in Pietermaritzburg, one year ago, I was excited to see the students were surfing the Web to prepare for a debate. But three months later I found out the school had cut their Internet service due to cost'.

Universal Service Agency co-ordinator, Kwa Zulu Natal

'Our school-based telecentres don't have a sustainability issue. The school is the core client; it is durable and networked to support. The telecentres are actually making a surplus: the issue is how to use it'.

Meddie Mayanja, World Links Uganda

The programme run by World Links for Development, an NGO sponsored by the World Bank, has provided schools with computer labs that must be open not only to students but also to the public, during certain hours. This model increases the range of people with access and also allows schools to recoup their costs. (See also Chapter 3.) Schools were informed of the programme and could choose to apply. Applications were accepted based on a feasibility assessment, including evidence that the school had the basic infrastructure (including electricity) and capacity to sustain the centre. Computer teachers were given short intensive training courses over school holiday periods, including one on technical maintenance which several people described as particularly valuable. Schools received funding for two to three years after their centres were set up, and by the end of that period were expected to be sustainable. They were also required to develop a business plan to help them achieve this purpose. Despite the serious setback caused by their ISP going bankrupt, the overall approach appeared to be working quite well. The national co-ordinator said many schools in the programme were reporting surpluses, and most had plans in place to upgrade and expand their equipment.

Some South African schools were also reportedly opening themselves to after-hours public access. No schools offering public access were identified within this research, although teachers in a number of schools with computer facilities said they were considering doing so. Security risks were a major deterrent to expanding access; schools were often victims of vandalism and theft, and computers were prime targets. Another major problem was the high cost of Internet connectivity, due to the high per-minute dial-up rates. Some schools had support getting onto the Internet, but were forced to disconnect because the bills were too expensive to pay. One innovative solution to this was being pioneered by Wizzy Digital Courier, the brainchild of Andy Rabagliati. This focused on partial connectivity, whereby the school's local area network (LAN) could be connected to the Internet via an automatic dial-up at night when rates were much cheaper, or by using a USB (Universal Serial Bus) flash drive, a miniature data storage device, couriered to and from an Internet connection that might be some distance away. Rabagliati, a programmer by training, had designed an open-source solution that allowed e-mails to be uploaded and downloaded this way and which also allowed teachers to identify web sites that would be automatically 'scooped' and downloaded onto the school's LAN. The project had been going for about eight months as of June 2003, and they had connected four schools. They were focusing most of their efforts in Eshowe, where they hoped to network all five schools using a hub system where the more wealthy former Model C School could provide some technical support to the others.[1] They found that besides the technology, project management and building up the internal capacity to use the equipment within the school took a lot of effort and it was worth going slowly. In Khayelitsha, for example, they had identified a number of schools in which computers were underused.

ICT access in libraries

'Some people are angry because of the few centres with these services . . . People want more centres close to them because they cannot afford to go to town. At the library there are few people because it is too expensive to use the computer'.

Research assistant's observations, Khayelitsha

In South Africa, a number of public libraries had been given computers and Internet connections with which they could provide public access for a fee. In Khayelitsha, only one of the five libraries within the township had Internet access. It had four computers that were given to the library in 2000 for the sole purpose of Internet use. Six library staff also received basic training in sending e-mail and surfing the Internet. However, most of the staff who received the training had since left. The City of Cape Town provided technical maintenance. Technical problems were frequent. At the time of research, only one of the four computers was working. From anecdotal evidence, this was a fairly typical experience. For example, another library in Bethlehem, South Africa, reported they had computers donated to them from a local college, but received no technical support and did not really know how to maintain them. This also meant that the quality of service tended to be low and users had to be independent.

ICT access in community centres

'[Setting up the centre] has taken a lot of networking ... You need to analyse your environment, decide what you need, and bring in expertise as necessary ... You need a bigger vision'.

Holly Luton-Nel, Director, Aleksan Kopano Resource
Centre, Alexandra Township

This model was quite common in South Africa. In Bhamshela, the telecentre had not been part of a community centre, but through the efforts of the tele-centre management, a building that housed various community activities had been built beside the telecentre along with a multipurpose community centre (MPCC) housing various government services.

One of the most influential role models for the GCIS MPCC programme was the Aleksan Kopano Community Resource Centre, opened in 1992 as Alexandra's first and only community centre. Alexandra is an old high-density township just outside of Johannesburg, with high unemployment rates and a reputation for violent crime. Aleksan Kopano Educational Trust is the not-for-profit body that owns the centre. While at first the centre served as a general purpose gap-filler in community needs, focusing on education, its purpose has shifted towards job creation, reflecting both an overall increase in community capacity over the years and the current priority needs of the community. The centre, a large three-storey complex, did not have a telecentre as such within it; rather, the various services one might expect to find in a telecentre were scattered throughout. A photocopier was at reception, Telkom public phone booths were on site, and the library had six computer stations available for basic off-line applications such as word processing, plus a separate room that operated as an Internet café. Both the off-line stations and the Internet café services were free of charge. In addition, it offered a subsidized six-month course that included computer training and basic business skills, designed to increase the employability of its graduates. Alex FM, the township's community radio station, was also a tenant of the centre. In this model, the ICTs available complemented the other services and activities at the centre, which was very active and dynamic. The other activities also helped to draw people to the centre. While all services were well used, they depended on outside support, mainly from the private sector. Its success in securing such funding was largely due to the dynamic efforts of the centre's director, Holly Luton-Nel. The Internet café was fully used when it was free, but when the centre tried to introduce user fees, people stopped coming. Since the centre could afford to offer only three hours of service a day to begin with, this created a potential problem for the sustainability of the café, Again, this was largely attributed to the high cost of Internet services. If and when better connectivity options were to become available, the resulting lower user fees might be more acceptable by the community. Nonetheless, because of high unemployment and poverty, people were generally reluctant to pay.

ICT access in NGOs

'I had thought that offering computer-related services would be income-generating. Now I question how this can happen because people here don't have money. Over the long term, we're taking a 30 per cent loss'.

Manager, KERIC

'RANET would be very expensive for one group to implement, but because of partnerships, the costs were minimal'.

Milton Waiswa, Manager, RANET Uganda

Some NGOs offered ICT access as part of their operations. Two very different examples from the case communities are the Khayelitsha Educational Resource and Information Centre (KERIC) in Khayelitsha, South Africa, and the RANET programme that was based out of a number of World Vision field offices in Uganda, including one in Bukinda, Kabale. The former was a stand-alone service that had been implemented on the initiative of KERIC's manager. The latter was an international programme implemented through a partnership among several large, high-capacity stakeholders.

KERIC, an NGO established in 1993, was located within a community centre complex in Site B of Khayelitsha Township. KERIC's aim was to contribute to community development by focusing on educational advocacy and lobbying, while developing strategies to improve local pass rates. It had offered a small typing pool within the office for the last four years. This included e-mail services and fax. Clients could ask the staff to use the computers, or use them directly. They were encouraged to do the latter, especially for preparing their personal CVs. However, since this office is also used for KERIC's other activities, high usage of these services leads to congestion. The typing pool was running at a loss of approximately 30 per cent and service prices were subsidized, since the actual cost was unaffordable. The manager said that although he had originally believed the typing pool would be an income source, the focus was on providing accessible service. This, however, posed some problems for sustainability, as ongoing donor funding for this kind of activity seemed unlikely.

RANET was an international programme centred on delivering seasonal climatic information to farmers, especially in drought-prone areas. In Uganda, the Department of Meteorology had partnered originally with World Vision Uganda, whose area development offices were situated in needy rural locations, equipped with knowledgeable staff, an administrative computer, and a source of electricity, often via a generator or solar panel. These offices served as the 10 original RANET sites (by 2004 these had expanded to 24 sites, through additional partnerships with ActionAid and Africa 2000 Networks). The existing computer was equipped with the satellite data receiver, which cost about US$50, and an antenna. It could download data from the World-space satellite, including the RANET climatic information and other data from the African Learning Channel. Dissemination of the information was through social structures such as farmer groups, public postings and announcements over local radio. Farmers were also provided training about how to use the climatic information to best advantage in planning their planting and harvesting activities, and this information was also disseminated largely through the farmer groups. One interesting outcome of this process is that the farmers wanted greater direct access to the information on the computers, and to a broader range of information. Thus, the programme was looking into acquiring dedicated computers for public use, which would begin to resemble a telecentre. However, no fees were in effect for these services.

The RANET programme did not attempt cost recovery through user fees. Rather, its running costs were kept low and mainly absorbed by the World Vision Area Development Programme (ADP) budget, since the computer, electricity, staff and transport (for technical repairs, etc.) were all supplied by World Vision. The additional equipment, an annual ream of paper and training sessions were paid for by the Ministry of Land and Water, which in turn received money from the RANET international programme. This international programme consisted of a consortium of various funding groups, most notably USAID. As RANET helped to achieve priority objectives of World Vision and the Ugandan government and appeared highly relevant to the communities it was serving, it seemed likely that they would continue to support the programme and that it would be feasible to do so within their main budgets.

However, lack of further resources limited the potential growth of the programme. It was up to each area office whether and how to disseminate information on topics other than climate. For example, the area offices also received a large amount of information on agriculture, health and water management. The RANET manager at the Department of Meteorology expected to use the same technique that he had used to date – namely, nurturing partnerships of mutual interest – to expand the programme.

Telecentres

South Africa

'We've started provincial forums to help telecentres. But almost all the tele-centre managers now are volunteers, and that means big turnover. I and one other are the only ones left in the province. Almost all the telecentres are not working, especially in Kwa Zulu Natal – that's our main problem, there's no money coming in. We don't even have enough to buy cleaning supplies. Some are thinking of stealing equipment because now all USA [Universal Service Agency] telecentres are insured'.

Telecentre manager, Kwa Zulu Natal

Both telecentres in the South African case studies were established through the Universal Service Agency. They were created by local NGOs that applied to the Agency after hearing about the programme. The telecentres were established as separate not-for-profit corporations, each governed by a board of directors. The Agency provided the initial equipment and training, while the founding partners supplied the rest. Although they were intended to offer a full range of services, including Internet access, most telecentres never achieved connectivity.

The Bhamshela telecentre, which began operations in 1998, was one of the earliest Universal Service Agency telecentres. It was set up in a rented four-room building near a taxi rank and equipped with a photocopier, five Pentium computers with internal modems, a printer, TV and video equipment, fax, an overhead projector and screen, and six Telkom landlines and handsets. It was one of the few telecentres that did get Internet services, which it offered for a year and a half until it ran into financial difficulties following a Telkom billing error, resulting in their phone lines being disconnected. The telecentre began to explore other service avenues, and began to offer computer training certified by a Durban-based college. Until this time, the telecentre had depended mainly on telephone services for its profit, while all other services, with the exception of photocopying, were hardly used.

The Khayelitsha Development Forum (KDF) established a telecentre in a primarily residential area of Khayelitsha in July 2000. In 2001, Microsoft donated nine computers to the telecentre. At the time of research, they had 10 personal computers and one staff person. The telecentre offered fax, computer, photocopying and printing services. There was also a pay phone nearby. The computer services included computer rental, typing and computer training. However, none of these had a big demand, and at the time, computer training sessions had not been offered for the previous four months due to the low demand. The telecentre was not breaking even. The manager had another job and subsidized the telecentre from this. He felt that marketing and the lack of Internet access were the main problems and he was taking steps to rectify both.

The experiences of these telecentres were unfortunately typical of the Universal Service Agency programme. Although the Agency had helped them get started (which had also been a slow process with frequent delays), the basic model was premised on the idea that they would then be able to sustain themselves through user fees. This turned out to be unrealistic for several reasons. Firstly, there were long delays in getting service. Secondly, the high costs of phone and Internet connectivity meant that centres often generated very little profit on these services. And thirdly, there was little ready demand for many of the services, and the centres did not have the capacity to build such demand. Maintenance was also an issue and problems could take months to resolve. According to research conducted up to 2000, about one-third of the Universal Service Agency's telecentres had managed to become sustainable by incorporating ICT-related value-added services for which sufficient demand existed (Benjamin, 2001a, b). But by 2003, some centres that had appeared stable a few years earlier had gone into decline. They had little financial capacity and so were unable to withstand large shocks such as equipment damage and theft. Chronic lack of finance had also limited the telecentres' capacity building and sometimes provoked internal conflicts. Some thefts appeared to be 'inside jobs'. After years of struggle, often without pay, many telecentre staff had become demoralized.

Uganda

'Looking at school-based telecentres, they're so much easier. They have security, management, more opportunities. Community-based telecentres have a kind of rigidity because of the low capacity and limited outlook of the management. Local government hasn't been that supportive'.

Former manager of Nakaseke Telecentre

'Telecentres work but people need to be assured and trained. They also need relevant content. There's a need for seed money in rural areas to grow demand'.

Goretti Zavuga, Council for Economic Empowerment
for Women in Africa [CEEWA]

All the telecentres included in the Ugandan case studies were funded primarily by Canada's International Development Research Centre (IDRC), and each was implemented by a separate agency. IDRC's Acacia programme had been one of the earliest and most prolific funders of telecentres in Uganda through a variety of different projects and partners.

The Nabweru telecentre, which opened in May 1999, was one of the first in Uganda and was administered by the Ugandan National Council for Science and Technology (UNCST). It was also ahead of the market in the subcounty, predating any similar public facilities. The telecentre was situated at the Nabweru subcounty headquarters, based on the rationale that the subcounty administration would be its anchor clients, and that they would play a role in its management and become its eventual owners.

Photocopying was originally the Nabweru telecentre's most popular service, followed by telephone. In the summer of 2001, the telecentre had an MTN pay phone installed outside, which became much more popular than the phone services within the building. When Internet access and e-mail were offered, they were provided by dial-up and cost Ush170 per minute. The telecentre was paying about Ush50 per minute for the connection and could not afford to charge less than Ush100 per minute. When cybercafés began to switch to wireless and bring down their prices to Ush50 per minute, the telecentre could no longer compete. It discontinued Internet services in March 2002.

Sustainability had always been an issue for the Nabweru telecentre. From its inception until August 2002, it was subsidized to cover expenses and all profits were banked. The financial management of the telecentre had been largely out of the hands of the local management. For example, they could not spend above a predetermined amount on each service, even if it would generate more profit to do so. The prescriptive structure of the budget also limited the telecentre's capacity to experiment with new or expanded services. After transfer, the local authorities were responsible for approving monthly budgets. In practice, while they continued to act as anchor clients for telecentre services, especially photocopying and typesetting, they had not otherwise taken an active interest in supporting and developing the telecentre.

The telecentre's monthly expenses were about 15 per cent short of its income. Customer numbers had dropped for phone and printing services because of competition, while computer rental services had always been in low demand. Computer training generated about half of its income, followed by typesetting and printing, the photocopier, telephone, consulting services to other agencies, and renting out the generator. The constant shortage of money limited many activities. For example, low local awareness was an acknowledged problem but there was no money to pursue an extended marketing campaign.

A number of new projects implied ongoing relations with external funders, which management welcomed as a needed source of support for the telecentre. One project was focused on documenting indigenous knowledge through video and written narratives. The World Bank supplied equipment for this initiative, and the UNESCO High Commission for Uganda led it. UNESCO had also contributed Ush5 million towards setting up a community radio station, and the subcounty Ush2.6 million. The radio station was due to start within a month or so. Staff members were also hopeful that the radio station would provide sufficient motivation and support to bring in a wireless Internet connection.

It is difficult to say whether local communication centres were inspired to begin business because of the telecentre. The out-of-the-way location of the telecentre probably limited its influence. Overall public awareness of the telecentre appeared low, even amongst those living nearby. Four independently owned phone bureaus were started in the immediate vicinity of the subcounty headquarters. These may have been influenced by the telecentre, but probably also by the moderate traffic of subcounty headquarters staff and visitors.

The Community Learning Centre in Lira was a Canadian Physicians for Aid and Relief (CPAR) pilot project funded by IDRC. The CLC was set up within the CPAR base camp on the edge of Lira town. Its core purpose was to train 200 women and youth in ICT skills in a manner that would improve their livelihoods, and to document the process. The targeted beneficiaries of the project were individuals who had been directly affected by the war, most of whom had been earlier abducted by the Lord's Resistance Army (LRA), and some of whom were orphans. Although the business plan included public access, not as its core activity but as a partial means of sustaining itself, this was not yet being offered; in fact, nothing other than some preliminary training activities had taken place. The project began in March 2002 with three years of assured funding, and took a number of years before this to develop. The slow progress was due in part to challenges in administration and management for both IDRC and CPAR. Finding and retaining capable staff were also problems. The centre had benefited from the involvement of two Engineers Without Borders interns from Canada: Ryan Sparkes and Avi Caplan, who between them provided about 10 months of voluntary support, focused mainly on technical issues and training staff. The Lira CLC was supposed to be one of four such centres, with the other three planned for more remote areas. However, the security situation had prevented development of the other centres, although one other outreach centre at CPAR's Loro base camp opened some months later.

The African Highlands Initiative (AHI) telecentres in Kabale were set up in 2000 as part of a larger initiative to improve the livelihoods of farmers in the African highlands and to help them redress problems of environmental degradation through participatory action research projects focused on improved farming methods and natural resource management. These tele-centres fit into this larger project as hubs for information generation, dis-semination and exchange. One telecentre was located in Kabale town and offered similar services to many of the commercial secretarial and computer training services. A second telecentre was set up in Rubaya at the subcounty headquarters. Rubaya is a mountainous area with poor roads and no electricity. The telecentre offered computer training, typesetting, photo-copying and phone services, and was the only place in the area that offered such services. It also had a meeting hall and a demonstration plot for experimentation with seed varieties and farming techniques. A library held a small collection related mainly to agricultural practices. It was within the vicinity of the trading area, four high schools, a polytechnic institute and a number of NGO offices. Although open to the public, its activities and structure specifically targeted farmers, schools, local NGOs and government workers. Both the Rubaya telecentre and the Kabale telecentre were translating brochures and pamphlets into local languages and making them available in the libraries. This service was free. They also planned to expand the collection to include textbooks and resources for students available to photocopy.

It had been expected that both telecentres would be connected to the Inter-net, but in 2003 neither of them had such access. The town telecentre had had dial-up connectivity that it disconnected when a nearby competitor (the Voice of Kigezi Internet Café) acquired a VSAT connection that provided faster, cheaper service. A wireless Internet connection had been planned for the Rubaya telecentre, but this proved technically unfeasible due to the mountainous terrain.

Both telecentres confronted major obstacles in becoming sustainable. The town telecentre faced strong competition and high overhead costs. It was not able to compete with commercial centres that were located in more visible areas, had greater flexibility to make immediate decisions about services and, in the case of the Kigezi Internet café, had a better Internet connection. The rural telecentre had no competition, but had to contend with poor infrastructure, high operating costs and a relatively low demand for services resulting from low awareness, literacy and income levels of the surrounding population. Equipment maintenance and the purchase of consumables, including telephone connection time, all had to be done via Kabale town and, for major repairs, Kampala. This could be a lengthy process leading to service interruptions of a month or more. The two full-time staff did not have the technical expertise or supplies to perform on-site maintenance and repair. The telecentre depended on a generator that charged batteries which, in turn, could supply power requirements for up to three days. Plans were made to switch to solar power as a more cost-efficient energy system.

The Kabale telecentre had been intended to be self-sufficient, and perhaps to contribute towards the running costs of its rural sister centre. This proved to be overly optimistic because of high competition, and it was only generating enough income to cover about one-third of its operating costs. The most serious problem, however, was due to an extended gap between periods of donor subsidy. Although IDRC had agreed in principle to fund the telecentres for another three years, the funds had been delayed for almost a year. This had serious consequences for both the centres in Kabale Town and Rubaya, limiting their ability to function and plan, and reducing staff morale. Staff members in Kabale Town felt their telecentre had the potential to be sustainable with greater up-front investment. In particular, acquiring a VSAT would allow it to compete and gain customers. Whether this would be sufficient to support its sister telecentre in Rubaya or allow it to focus on its value-added development objectives was questionable: the necessity to survive and the strategies staff had identified for doing so were quite unrelated to development-focused activities, which tended not to be income-generating.

The end of the first three-year phase of donor funding brought AHI to the conclusion that rural telecentres and information provision required subsidization, because the target users simply could not afford to pay and did not attach monetary value to information.

Telecentres and the local market

'At first there was some lack in the computer field in the telecentre – a lack of promotion of computer services because the telecentre was busy with the telephone. But if you want your thing to work, you need to strategize. Awareness about computers has changed, the staff have gained experience in computers and now they are counsellors to others'.

Telecentre board member, Bhamshela

Telecentres and other non-market ICT ventures are sometimes observed, as well as expected, to develop a local market for other public ICT services by creating local awareness and demand for such services. In South Africa and Uganda, the following relationships between non-market initiatives and local market activity in ICT provision were observed:

- Some places simply do not have the capacity to sustain a market. It may also be quite difficult to establish a funded project there due to lack of infrastructure, capacity and isolation. In those instances where a telecentre or other funded ICT project can be established under such conditions, it will have an effective monopoly. However, it may not be able to charge more than minimal prices, insufficient to cover costs, and will not be able to create a market until other conditions, such as infrastructure, also become more favourable. The biggest contribution that a telecentre or funded project can make toward establishing favorable conditions is often to increase the awareness and capacity of the staff, who may then go on to create other services.
- Some places already had an active market in ICT services, if not at the time the telecentre was first planned, then by the time it was implemented. Internationally funded projects, especially those involving multiple partners, can be complex and slow, whereas local entrepreneurs can move quickly. In such cases, it is possible that the telecentre has further strengthened the perceived viability of ICT services amongst local entrepreneurs, while at the same time, creating competition and making it more difficult for others to enter the market and gain a profit.
- In some places, the impact of telecentres on market activity has been minimal, even where they were early on the scene. This was due to internal limitations affecting services, to poor marketing or to poor location.
- In places such as Bhamshela, where infrastructure was reasonable and there was already a strong local demand for some ICT services such as phone, the telecentre raised awareness within the small business community about the opportunities for ICT provision, and acted as a catalyst for such services. More unusually, the MPCC provided an anchor market for ICT services not yet available, and prompted local business to start offering them.

Part of an integrated development strategy

'Telecentres can't stand alone in Africa – they have to work in networks and partnerships'.

Former manager of Nakaseke Telecentre

Both South Africa and Uganda were in the process of implementing major rural development strategies in which ICTs had the potential to play a key role.

Multipurpose Community Centres (MPCCs), South Africa

'The MPCC is a means of integration, the "mother of all". The NCRF is positioning itself around the MPCC, but we need to safeguard the independence of community radio'.

Chris Armstrong, National Community Radio Forum (NCRF)

In South Africa, discussion and brainstorming in the mid-1990s about the role of ICTs in furthering development in rural and township areas produced a number of ideas, including the multipurpose community centre, or MPCC. The government and civil society organizations collaborated through a number of forums in developing the notion, one of the earliest being the Empowering Communities in the Information Society Conference, held at Helderfontein in 1996. The government perspective has tended to focus on the MPCC as a vehicle for government service delivery plus some two-way interaction between government and citizens. In contrast, the South African NGO sector has emphasized the MPCC as a tool for community organizing, development and creating local media.

Initial research in 1998 attempted to catalog what sorts of existing organizations might be considered MPCCs, using the following definition: 'An MPCC is an organisation offering a range of developmental services (including information services) to a specific community and with a large degree of community involvement' (Benjamin, 1998). The research identified 235 centres fitting this definition, although many of these had few or no ICT services. Implementation of government-sponsored MPCCs began formally in 1999 when the Government Communication and Information System acted as the secretariat of a large multi-stakeholder process operating through national, provincial and municipal intersectoral committees and working groups. While in practice many MPCCs, especially in rural areas, remain without connectivity and some do not have electricity, the idea is that they should be fully connected to ICTs by containing a Universal Service Agency telecentre as one element, both to support their own services and for public access.

At the time of research, about 30 MPCCs had been established across the country and the others were expected to be established within the next year. An MPCC had recently opened in Bhamshela, thanks largely to the efforts of the telecentre's management. It was owned by the municipal authorities and began operations in early 2003, housing six government offices from the departments of Home Affairs, Labour, Education, Health, Agriculture and GCIS which offered a variety of on-site and outreach services. Those offices housed in the MPCC had signed an agreement that they would not bring their own equipment with them (photocopiers, printers and fax machines), but would use telecentre services, thus becoming anchor clients. By the time the MPCC opened, the telecentre could no longer offer these services. Motivated in part by the urgent needs of these government offices, the nearby Village Bank began to offer photocopying and fax services and the telecentre referred customers to them. Thus, the MPCC in this situation had helped to stimulate the growth of ICT services without actually providing them. While ICTs could help the functioning of MPCCs, which in turn had the potential to fulfil a variety of community needs, the biggest obstacle, common to all ICT-related endeavours, was the high cost of landline phone and Internet services, meaning that many centres remained without.

National Agricultural Advisory Service (NAADS), Uganda

The National Agricultural Advisory Service (NAADS), which replaced the old government extension system, has a communication strategy that draws on a variety of techniques and media. Telecom tools are one component of this strategy. In some cases, NAADS was already working with telecentres at the time of research. Access to accurate and timely market information had been identified as a priority area for NAADS.

NAADS focuses on the 'demand side' of agricultural extension services by organizing farmer groups that can collectively demand and buy information related to agriculture and markets. NAADS helps the groups to pay in the beginning, on the assumption that once they are putting new farm practices and market knowledge into practice, they will be successful enough to buy the information themselves, while the collective activity reduces the per-farmer cost. Thus, NAADS is helping to create effective demand for the kind of services that telecentres provide.

The rural telecentre in Rubaya had been working with NAADS since 2001, since they shared many goals. The telecentre staff had been contracted by NAADS to help organize local farmer groups. This relationship appeared to be of great mutual benefit. It might increase the telecentre's value to the community, if the NAADS farmer groups learned to seek and apply the different kinds of information the telecentre might provide. In Nabweru, NAADS was also beginning activities and thus opened up the possibility of collaborating with the telecentre there, with similar potential. Since both of these telecentres had struggled to increase demand for their agricultural information services, the presence of NAADS could potentially help to sustain and increase the value of these types of services. The co-ordinator of RANET Uganda was also in dis-cussion with NAADS about possible collaboration. Ideally, this programme, which was far-reaching and integrated into overall national development efforts, might act as an umbrella for many otherwise disparate projects and programmes aimed at providing development-related information in rural areas.

Analysis and key lessons

This chapter has so far reviewed the various types of ICT access found in five case communities (and a few other places) in South Africa and Uganda, how these centres began, and how they have continued. The chapter's concluding section explores a number of broad themes emerging from this study.

Firstly, telephone access appears much easier to establish and maintain than other forms of ICT access. For this reason, it is worth distinguishing between phone and other ICTs when one considers universal access.

Many of the challenges related to centre start-up and sustainability relate to the local market. Particularly, the relationship between the market and various initiatives established outside the market is an important but complex one. The relationship between local access centres and national operators is always an uneven one – the latter have much greater resources and political leverage than the former. Yet local access centres, whether public or private, provide value to the communities they operate in, to national development, and to the major operators. Their contribution needs to be recognized by regulators and other bodies so that they do not constantly have to struggle while large operators reap most of the profits.

Sustainability is often a challenge for both for-profit and not-for-profit access centres. While there is no easy or magic solution to the problem of providing long-term, far-reaching ICT services to populations who often lack the means to pay for it, some strategies can reduce the cost, complexity and associated risks of providing ICT access.

Finally, if access centres are to make a significant contribution to universal access and national development, they need to be integrated into larger strategies. The market alone will not create access in all parts of either country. Even where centres are created with external funding, they are generally not sufficient to build local demand. These strategies are especially important in rural areas as well as for reaching poor urban populations.

Telephone, an ICT with ready demand

In the 1990s, strategies for providing Internet access were often conceptualized as extensions of strategies for providing telephone access. At that time, this made sense since, for technical reasons, policy makers intended to achieve universal access goals primarily through fixed line telephones, and cellular technology was perceived as an elite service. The concepts of multi-purpose centres and one-stop shopping for communication services – that is, providing a full range of services using the same infrastructure, building and staff – were seen as ways to achieve economies of scale.

In both South Africa and Uganda, cellular phone subscription has greatly outstripped the number of landline phones. This is even more marked in rural and low-income areas where cellular phones are often the only form of telephony available. But cellular technology is not well suited to data transmission and does not readily support Web browsing. Although Web browsing by cell phone is available in South Africa, it is expensive and few people use it. Thus, the technical and infrastructural link between telephone provision and Internet provision is not as strong and straight-forward as it once appeared.

The other important difference between Internet and telephone access, consistent across all the case studies within this research, is local demand. With telephone services, uptake is quick. And when people have money, they are willing to pay, especially if they can purchase by the unit without large costs. Conversely, there is little upfront demand for Internet access, although this demand was more evident in Uganda, possibly because of the larger proportion of people with family and friends overseas.

The cost of Internet access, coupled with low awareness of the technology and little expertise in its use, are the main factors preventing people from using it. Those with little or no experience with computers may quickly catch on to the basics of Web browsing, just as they are able to master options on their cellular phones. Younger people, in their teens and twenties, are generally quicker and more interested in learning than their elders. However, typing on a keyboard is time-consuming, and per-minute pay schemes mean that people without any typing skills or basic familiarity with the layout of keys will pay considerably more because of their own inefficiency. This is a powerful practical disincentive to learn, implying that those who are most likely to use public pay-as-you-go Internet access have other places, such as a university computer lab or workplace, where they can at least master some basic computer skills. This further implies that this type of public Internet access largely fails to serve those without other options, and is not, by itself, a viable means of providing universal access.

The same methods of pricing and delivery that seem to work well for telephone do not seem to work well for Internet access. Early multipurpose centres offering a range of services have tended in practice to focus upon one or two popular services, neglecting those for which there is little ready demand and limited staff capacity to support. These underutilized services do not add value to what the centre does; rather, they burden the centre and its customers with added costs.

Access centres and local markets

Shared access ICT centres have a number of ongoing relationships with the local market to sustain themselves. Most not-for-profit initiatives still depend to some extent on generating revenue from their customers for financial viability, and this involves certain assumptions about what the local market can support. Seven types of intervention to establish access centres are summarized in Table 4.3. Each intervention type depends on a different type of relationship to the local market to be successful. The last two are most suited to rural and traditionally underserved areas and populations.

The first strategy is 'hands off' (no intervention): a centre is both established and maintained by the market. In cities, small towns and their peripheries, local entrepreneurs perceive market opportunities and set up their own access centres, often motivated by services they have seen elsewhere.

The second intervention type assumes that the local market can support access services but that no local entrepreneurs have so far been inspired to do

Table 4.3 Interventions to create access and intended market response

Intervention	Intended market response
1. None	The market will establish and maintain appropriate services.
2. Demonstration project	The market will duplicate and sustain additional services in response to user demand shown for the demonstration.
3. Start-up grant or support	The market will maintain the services created through start-up grants.
4. Establish services and build the market (through short-term subsidy)	As demand increases, the market will maintain or duplicate the services with greater efficiency, eliminating the need for subsidy.
5. Ongoing subsidy (to operator or to end user)	The market will meet the gap between subsidy and actual price (through user fees).
6. Build an anchor market	Anchor users will cover a proportion of running costs and may create value-added public services (eg, MPCCs). The market will meet the gap between anchor income and running costs.
7. Embed new services in existing structures	The host structure will absorb some costs and may also be an anchor user; the market will cover additional running costs of providing the public service.

so. If people see from a demonstration project that they can start up an access centre and run it profitably, they will do so. A number of early communication centres in Lira claimed that their success quickly inspired others to follow suit. In Bhamshela, the telecentre demonstrated strong local demand for telephone services.

The third intervention addresses a situation in which local entrepreneurs or NGOs may not be able to start up an access centre, even if they wish to and there is ready local demand, because they lack capital and technical training. Support to meet such requirements allows the centres to start up successfully and then run sustainably on their own or even generate a profit. This was the scenario assumed by the Universal Service Agency telecentre model. Unfortunately, in many communities, things did not occur according to plan. The multiple services offered by the telecentres actually weakened them since the few services for which ready demand did exist had to subsidize the rest. This model also assumes that service of reasonable quality and at reasonable cost is available from national operators.

Intervention 4 recognizes that there may not be ready demand for ICT services amongst the local population, usually due to lack of previous exposure. Also, it may take some time for the centre to establish itself and build up its internal capacity, especially since the staff may have limited prior experience with ICTs. A short-term subsidy usually covers setup and running costs for about three years. This was the approach taken for many of the funded telecentres in Uganda. While sound in theory, many technical difficulties tend to arise in practice, especially with infrastructure, and the funding must be flexible to allow the centre to respond fully to areas of strong local demand. Even as a market does develop in certain services, private operators may prove more nimble and compete successfully with the centre.

Intervention 5 recognizes there may be a basic gap between the cost of providing service and the required number of clients able to pay. Ongoing subsidization makes up the difference. This often happens when local NGOs provide services, usually not by design but because demand and willingness to pay proves lower than anticipated. The problem with such subsidization is that it is difficult to secure over the long term and tends to be unreliable. Thus, issues of affordability need to be met either at the national level by lowering the cost of basic services, or the centre itself must find a way to reduce the cost of offering the service.

Intervention 6 is well suited to rural areas where ability to pay is low. However, it still requires the presence of affordable, reliable basic telecom and electrical infrastructure. An anchor user, such as a government office, covers some proportion of the operational costs of the services, and may even create extra public services such as training or access to specialized information. The public can use the services as well, for an affordable fee. The combined income sustains the centre.

Finally, in the seventh type of intervention, the anchor user owns and hosts the centre within its own infrastructure. The World Links school-based telecentres are a good example of this. The host structure absorbs some costs and usually benefits from using the services itself, while public user fees help to maintain the facility.

Sustainability inside and outside the market

The sustainability of access centres has been an ongoing concern, typically focused on the issue of direct revenue generation through user fees. Such fees often undermine efforts to increase the reach of access centres, especially amongst populations with no previous experience of ICTs.

In practice, sustainability can be thought of as a strategic management issue that aims to reduce costs and to reduce or manage complexity. For example, building a telecentre as an institution from scratch involves great complexity. Embedding ICT services in existing institutions and institutional practices simplifies matters. A school-based telecentre is one way to do this; but it does imply an expansion of the school's functions and structures, as well as a new relationship between school and community. RANET is another example; in this instance, climate information was delivered via the existing World Vision programme structures and staff. This sort of approach also allows for costs to be at least partially absorbed by the host institute, especially where the ICT services also add value or support existing functions. Such institutions can also act as anchor markets for services, which is the cornerstone of the MPCC/telecentre relationship in South Africa.

National and local operators

A key problem facing many access centres is the high cost of telecom services available to them, often accompanied by poor service. For instance, complaints about billing errors were common in both countries, as were complaints of low-quality service, particularly the poor line quality on long-distance dial-up. In both countries, areas outside of major population centres may still have only one or two options in terms of both operators and services, and little room to bargain. For telecom providers, access centres bundle local demand to make it more cost effective to deliver services to low-income populations. They also increase both people's exposure to services and their overall use of telephone networks since access centre users call to, and sometimes receive from, private lines. In other words, telecom operators benefit from access centres, even if such centres and populations are not the most lucrative segment of their business. Yet because access centres are small and often far from the capital cities, they have relatively little economic or political power. In Bhamshela, for example, the Telkom container was well used, but almost all of the considerable profit it generated went to Telkom, even though the enterprise was started and nurtured by a local person. One simple but important way for regulators to pursue universal access, therefore, is to put in place regulations that explicitly recognize the value of local access centres, both public and private, and give such centres preferential rates consistent with a reasonable profit margin. Vodacom has done this with its phone shops. Although motivated by licensing obligations, this programme is now self-sustaining and has also benefited Vodacom, mainly through increased promotion of its brand.

Access centres in the context of larger strategies

Dymond and Oestmann (2002) argue that the 'access gap' is really two gaps. The first is that between current and potential reach of the market. The second is the gap between the potential reach of the market and universal access. That is, there are certain population segments that will not be able to reach and afford ICT services even when the market is operating under ideal conditions of full and fair competition. Given, however, that the market is not operating under those ideal conditions, governments, and especially regulators, should focus on stimulating market growth, since this strategy is cheaper and more sustainable than massive direct subsidization.

This research has argued that there is a qualitative difference between the access gap in phone services and the access gap in Internet services. This difference implies the need to address each of these situations with a different strategy, where doing so has been identified as part of the public interest. Melody (1998) has argued that infrastructure provision on the supply side does not address lack of demand. From this research, it would appear that his claim applies much more strongly to Internet access and computer services than to the telephone. Because user fees strongly inhibit the creation of new demand through access centres, creative ways of establishing ICT access through low- or no-fee services are necessary. RANET is one example of this approach. Otherwise, shared access centres are quite limited in creating universal or even popular access in the absence of other conditions such as universal secondary education where all schools are equipped with ICTs. The majority of public attention and funds should therefore be directed towards these more fundamental requirements, while access centres should be left to the market. Arguably, this is what is happening in urban areas where the market does support access centres such as cybercafés.

Urban cybercafés service mainly those people who also have access elsewhere. Large segments of urban populations, notably those with low levels of education and professional skills, remain largely untouched by these services. If universal access and inclusiveness are genuine policy goals, these populations must be targeted by other means and strategies. Increased access to formal education that incorporates ICT use and training, including adult education, would seem to be the first option. Continuing to stimulate the market to offer access centres in urban areas remains a useful policy tool, since it does not cost the public anything and there are potential benefits from doing so. However, achieving universal access is not likely to be one of these benefits. Rather, the presence of such centres in the absence of larger strategies to promote digital inclusion implies a worsening local digital divide.

In most rural situations, the market is a limited force. With the notable exception of some cell phone penetration, it is unlikely to generate access to ICTs spontaneously and cannot sustain such access. Initial strategies to provide access to ICTs in rural areas need to be carefully integrated into other development efforts. This helps to minimize costs and to provide tangible benefits to local people without those same people needing, at the outset, to pay for services, know how to use them, or even know much about them at all. Such a strategy can eventually stimulate greater awareness and greater interest in further exploration. RANET is a good example of this. Over time, and depending upon other conditions, this can help to prime the market so that ICT services can eventually be supported.

CHAPTER FIVE
Local livelihoods, reach and development impact

Livelihood strategies are methods people use to survive and thrive. People choose their strategies according to the surrounding conditions, including risks, opportunities and available assets.

In Chapter 4 we examined how different types of access centres start up and survive under various conditions. In this chapter, we take a closer look at who uses these centres – and ICTs more generally – and for what purposes. A basic understanding of common livelihood strategies frames an analysis of both motivations for using ICTs (eg, they are a perceived necessity within a chosen strategy) and barriers to their use (eg, an ICT training course is too expensive). This same understanding can stimulate fruitful speculation about the likely contribution of ICT use to the well-being of an individual, a family or the overall community.

The first section of this chapter describes the types of livelihood strategies used by people across the case studies. The second explores usage patterns that typify each of the access centre types introduced in Chapter 4, drawing on the livelihoods framework and evidence from the case studies. Based on this, the final analysis and policy recommendations focus on how to maximize the potential development impact of access centres and strategies.

Livelihoods and priorities in the community cases

Living conditions varied considerably within and across the five community case studies, depending on the threats people faced and the assets available to them. People in rural areas typically depended on the natural resources around them for much of their livelihood, but also on some sources of monetary income, which provided added security in times of food shortage or sickness and allowed parents to pay school fees. People in urban areas had more opportunities to obtain money but little access to natural resources. Family and friends were an important resource for most people. Sharing assets and burdens within the extended family was a strategy for coping with poverty, poor harvests, unemployment, displacement, sickness and death in the family.

Common strategies

Although each person's situation and choices are unique, this research groups livelihood strategies into five broad types: survival, subsistence, subsistence plus, self-employment and formal employment. Two cross-cutting strategies, that is, used by all people except those in 'survival mode', regardless of life-style, were the acquisition of assets and the maintenance and expansion of family and social networks.

Survival
Survival is a short-term, reactive strategy, usually triggered by calamity, whether it be personal, environmental or political. In Lira, the camps for internally displaced persons are full of those in survival mode. These people are not in a position to plan much beyond the immediate requirements of the day. Other examples are the sick, the injured and children whose guardians have died.

Subsistence
A subsistence livelihood depends more on natural assets than a cash economy. People living a subsistence lifestyle are typically small-scale farmers who grow crops and rear livestock. They also use traditional practices and natural, locally available materials for building houses, getting drinking water and fuel, and treating illness and injury.

Subsistence living is vulnerable to climatic conditions such as drought and unseasonable variations in temperature, which have been occurring more frequently in recent years. Political conflict and coercion over land, as well as environmental problems such as soil erosion and deforestation, also threaten those living at subsistence level.

Subsistence plus
Money is increasingly required and valued by households for the opportunities it can bring. School fees are the single largest expense faced by households that depend largely on subsistence strategies. While both South Africa and Uganda have minimal or no official school fees for primary education, there are often hidden fees or other associated expenses, such as school uniforms and books, plus the opportunity cost of sending children to school rather than keeping them home to help with work. Many parents also hope to provide their children with an education beyond primary school, which increases the overall expense. Having money is also very useful for dealing with the shocks that subsistence living is so vulnerable to, such as illness and seasonal food shortages, plus the expenses associated with weddings and funerals.

People in subsistence lifestyles typically try to maintain some assets that they can sell if need arises, often in the form of farm produce or livestock. Otherwise, they may manage to secure some income through a small side business, part-time or casual work, or remittances from a family member.

Self-employment

Business conducted through self-employment is a common livelihood option in and around Ugandan towns and, to a lesser extent, in rural areas. It often supplements subsistence farming. In Uganda, lack of formal job opportunities necessitates this approach. For women, small business provides a degree of control, allowing them to balance income generation with housework and childcare. However, in many instances self-employed women do not aspire to expand their businesses because of the demands of their home-related duties.

Low literacy and numeracy can make small business risky and most entrepreneurs do not seem to base their economic activities on any kind of market research. This was certainly evident in the ICT sector at least, where people often based their business strategies on a combination of copying others and optimism.

In South Africa, people generally have a less entrepreneurial attitude and tend to look first for formal employment, including casual or untrained labour. However, people living in Khayelitsha may have other relatives depending upon them and the unemployment rate is high. They typically turn to small business to make a living if they cannot find a job.

In both countries, obtaining start-up capital and capital for expansion is difficult. Bank loans are not popular because interest rates are high, especially in Uganda. People typically use their own savings or loans from family members. Lack of capital prevents many people from beginning businesses or restricts the type of business they can do. It may also limit the growth of businesses that do get started, due to perpetual underfunding.

Formal employment

Jobs in the formal sector are difficult to get because there are so few of them and competition is intense. This is especially true in small towns and rural areas, where such opportunities, normally rare, tend to be with NGOs. Otherwise, people must relocate to larger cities. Formal sector jobs typically require at least secondary or some post-secondary education. Families with the necessary funds often invest in the advanced education of at least one child, hoping this will eventually lead to formal sector employment. In turn, those who eventually get such employment usually bear responsibility for a lot of their family's economic needs. Having a few salaried workers within a large extended family can dramatically improve the living conditions of the whole family. The situation in South Africa differs from that in Uganda because there are more opportunities for salaried, regulated work, including domestic and farm labour. These types of employment often require the workers to travel far from their homes. Some workers, especially skilled workers in Uganda, venture out of the country to seek better employment options. The remittances they send back can become the most important income source for their relatives in Uganda.

Cross-cutting strategies

People trying merely to survive are rarely in a position to acquire assets or expand social bonds. But in all other categories, these substrategies feature prominently in people's lifestyles – and with good reason. Family bonds have been especially important in allowing family members to withstand shocks. As a result of various social, economic and political crises, however, family bonds have come under immense strain. While family is traditionally important in both countries, Zulu and Xhosa cultures especially (in Bhamshela and Khayelitsha respectively) put great emphasis on the importance of extended family.

Acquiring assets

Whether people were engaged in a primarily subsistence or cash-based life-style, investing in and acquiring assets was a basic strategy commonly used the across the case studies. Acquiring assets normally implies some investment, either time or money, which is expected to provide a pay-off in the form of increased security, income opportunities, or both. What types of assets were most valued and how to acquire them depended on the specific situation of the respondent. A house was the asset that most people aspired to own, whether they built it themselves or purchased it. Farmers were typically inter-ested in acquiring land and livestock. Means of transportation were also commonly desired, especially cars. In all cases, education was seen as an increasingly necessary and important asset for improving the future of oneself and one's family.

Maintaining and expanding strong extended family bonds

Relationships with family, friends and acquaintances are another type of asset that was particularly important and productive in all the communities surveyed. Of all the case studies, people living in Bhamshela and Khayelitsha put most emphasis on the importance of family, probably for cultural reasons. In all places, family had traditionally served important economic as well as social functions, and the importance of these had if anything increased under modern pressures. Rural people especially looked to family members to venture to towns and cities and share the income they were able to generate there. South Africa traditionally has a high level of internal migration. During apartheid, only men could get passes to travel to cities where they could find work and send money back to their rural relatives. Although times have changed and whole families now move wherever they like, the pattern of urban-rural links remains strong. In Uganda, those who can have often chosen to leave the country to seek opportunities in places with better economies. These days, some people from neighbouring countries prefer to seek work in Uganda; but Diaspora links continue to be important.

All five communities had been visibly affected by the HIV/AIDS epidemic. Many people had to care for additional dependent family members. This, combined with high unemployment and limited options for income generation, put great demands on those people who did hold such positions. Besides providing economic support, especially for payment of school fees, urban-based relatives were typically expected to mentor and sometimes host younger family members so that they too could find work.

Local priorities

Table 5.1 shows the top priorities named by respondents in each household survey. Respondents were asked to name their main priorities in an open-ended manner (ie, they were not given a list of categories to select from), and their responses were then categorized. These responses give some indication as to the types of livelihood strategies most commonly employed in each case study. Education and housing were the top priorities named by all respondents. Many responses were specific to the particular situation (time and place) of the survey participants. For example, people in Lira focused mainly on basic survival. Their concerns were food, health and security. Their educational aspirations also tended to be modest – they wanted to be able to send their children to school, but were hampered by required school fees, and sometimes by insecurity and forced displacement.

Table 5.1 Top priorities named by household respondents

Priority	Nabweru	Bhamshela	Khayelitsha	Lira	Kabale
Education (for self, children)	25%	39%	39%	30%	67%
Build or buy house	29%	58%	41%	12%	18%
Start or expand own business	40%	6%	20%	0%	38%
Money	9%	28%	13%	39%	14%
Buy car or motorcycle	29%	32%	22%	0%	7%
Find a job	10%	22%	42%	0%	10%
Health	0%	0%	0%	41%	5%
Food	0%	0%	0%	44%	0%
Farming	12%	0%	0%	10%	17%
Family	0%	15%	13%	0%	0%
Marriage	0%	9%	0%	0%	19%
Security/peace	0%	0%	0%	26%	0%
Travel abroad	12%	0%	0%	0%	6%
Caring for children (own & orphans)	0%	0%	1%	18%	0%
Buy land	17%	0%	0%	0%	0%
Learn computers	0%	0%	7%	0%	8%
Become a professional	0%	0%	12%	0%	0%
Bicycle	0%	0%	0%	10%	0%
Religion	0%	0%	0%	8%	0%
Radio	0%	0%	0%	8%	0%
Clothes	0%	0%	0%	8%	0%

ICT access and use in the case communities

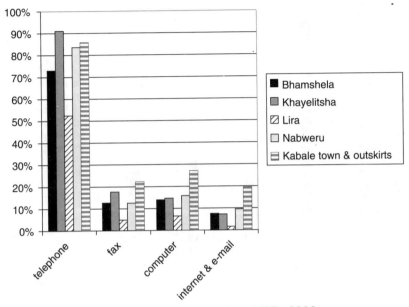

Figure 5.1 Percentage of respondents who had used ICTs, 2003

Figure 5.1 shows the proportion of household survey respondents, by community, who reported having used phone, fax, computer and Internet/e-mail. Not surprisingly, Khayelitsha holds first place in phone use since it is a densely populated urban area in the wealthier country. More surprising is that Kabale holds second place for phone use and first place in the use of every other ICT. For Internet and e-mail use, respondents in Kabale reported more than double the use rate of any other community.

The people included in this sample in Kabale are those from the town, who were mainly sampled from a relatively wealthy suburb called Kigongi, and those from surrounding areas, mainly sampled from a trading post about seven kilometres from town. In these urban pockets, awareness and use of ICTs appear relatively high, whereas those households sampled from the rural areas, in Bukinda and Rubaya, had very low levels of use. This is illustrated in Figure 5.2, where data from town, outskirts and rural respondents are disaggregated, further highlighting the urban-rural difference. The stark urban-rural difference is largely due to differences in infrastructure.

Both Uganda and South Africa have rural electrification programmes in place that will make the spread of these technologies much easier.

Radio is the one technology that was widely prevalent in all areas under study, and which significantly narrowed the rural-urban divide. Bhamshela and Khayelitsha were both within broadcast areas of community radio

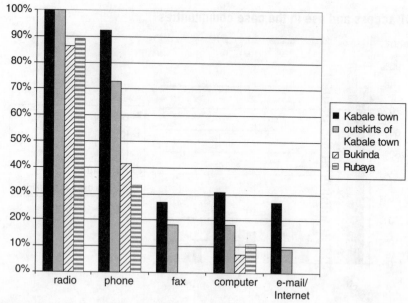

Figure 5.2 Percentage of respondents who had used ICTs, Kabale District, 2003

stations, while airwave liberalization has resulted in a proliferation of local FM stations in Uganda, providing most listeners with a choice of at least three or four. People living in camps for internally displaced persons and who were from the more remote parts of rural northern Uganda were the one group where radio use seemed to be less common. In rural areas, replacing batteries can be a problem. Frequently, a rural household will have only one radio and the man of the house will dominate its use. Nonetheless, both genders in all areas reported listening to the radio on a daily or weekly basis and using it as one of their major sources of information. Besides being information sources in their own right, radio stations, in Uganda especially, often associate themselves with and promote communication centres, Internet cafés and ICTs generally.

Fax, in contrast to radio, was not a particularly widespread or valued ICT in any of the case studies, and seemed unlikely to become so, since it compared unfavourably with alternatives such as e-mail for most purposes. It was not used by many people in any of the case studies, and those who used it, did so less than once a month. Uses were most often related to the formal workplace, sending in job applications, or to interactions with formal institutions such as banks, government offices and educational institutions. A few people in a number of different places (Bhamshela, Lira and Kabale) also mentioned using fax as a cheaper alternative to phone calls when maintaining contact with family overseas or internationally.

Phone shops

'I would use a phone to fit into the community and development'.

Household survey respondent, Rubaya, Kabale

Phone use has increased dramatically across all communities included in this research (see Figure 5.3), mainly because of the expansion of GSM cellular

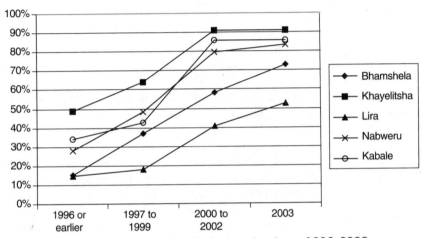

Figure 5.3 Percentage of population who had used a phone, 1996–2003

networks. Public fixed wireless phones, offered through public phone booths and franchises set up as part of operator licensing agreements, and through privately owned phone shops, have closely followed the expansion of these networks. While private cellular phone use has been much more widespread than generally predicted, public access through fixed wireless is the most common and most affordable type of use for most people.

Phone shops are more prevalent and tend to be more popular than phone booths, which can often be out-of-service and require prepaid cards. Phone shops also have attendants who can provide customer assistance. In many phone shops, they will dial the number for the customer so that all the person needs to do is speak into the receiver. Phone shop attendants in Lira reported that not everyone had yet mastered this; some customers even tried to put the handset on their heads.

Phone shops are used by all kinds of people, although most customers live within 20 minutes of the service. Thus, the biggest obstacle to use is for those living in rural areas where low population density, lack of electricity, and sometimes lack of a GSM network prevent phone shops from operating. Where a network is present, some rural people, especially those who travel to town regularly, will have cellular phones and often sell air time to others.

Phone shops do not offer all the benefits of private cellular phone use. While some phone shops, especially in residential areas, will receive incoming calls, many do not, and certainly it is not as convenient as directly calling some-body. Many phone shops also do not offer the same level of privacy as a personal phone.

The single largest use of the telephone, across all case studies, was social communication – firstly with family members and secondly with friends. The telephone was noted as an important communication tool for maintaining ties locally, nationally and internationally. In Bhamshela, people reported making calls specifically to request support and to announce deaths. Likewise, people in Lira, who tended to use the phone less frequently, did so to request support and to report emergencies. Those in Khayelitsha reported the highest frequency of calls related to family (86 per cent), and used them to maintain links with relatives in rural areas whom they would sometimes support eco-nomically as well.

Business-related use was the second most prominent. It was reported in all communities, although the proportion of phone users who made use of the telephone for business ranged considerably, peaking at 52 per cent in Khaye-litsha, but with only 12 per cent in Bhamshela. In Khayelitsha, people used the phone most often in formal employment, and also for informing their employer when they were sick. In Lira, Nabweru and Kabale, small businesses employed the phone in co-ordinating trade and co-operation between them-selves and businesses in other regions, especially Kampala. Cellular phones also allowed people to receive calls from potential customers. For example, some taxi drivers in Kabale and Khayelitsha carried them and could give out their numbers to potential clients.

Other reasons for using the phone varied. Emergency use, especially related to health, was particularly valued in Lira, but also in other places. In Lira, almost one-quarter of respondents recalled using the phone to make inquiries regarding school fees and educational programmes. People in Kabale also mentioned this.

Provision of telephone services was an important source of employment across all of the case studies, except perhaps in Bhamshela where there were only three jobs at the two phone shops. In other communities, entrepreneurs had found that the high level of demand for service and relatively low over-heads could sustain many small businesses. The most profitable were those with a Vodacom franchise, such as the one shown in Photo 5.1, since they got preferential rates from Vodacom, thus guaranteeing a generous margin despite heavy competition.

In both South Africa and Uganda, a number of innovative services providing information through Short Message Service existed at the time of research. For example, Foodnet and Gordon Bell (owner of Radio Lira) had come up with a project to allow people to obtain commodity prices via SMS. The user would 'SMS' the name of the commodity (for example, 'cassava') to the service's number; in reply, he or she would receive a list of up-to-date market prices in major trading centres across the country. However, during the course of the research no one who was interviewed mentioned using such services.[1] While

Photo 5.1 Vodacom container in Bhamshela

this can be partially explained by the fact that many of these services were new and people did not know about them, there did not seem to be any immediate demand for information-driven services.

Internet cafés

Internet cafés were the main venue of Internet use in the cases surveyed. This was because in most areas, especially in Uganda, relatively few work places had Internet access, even if they had computers, and no one reported a home Internet connection. In Bhamshela and Khayelitsha, where no Internet cafés and few other public Internet sites existed, Internet use was also very low (see Figure 5.4).

In Lira, where awareness and use were most limited, some of the first and most dedicated Internet users were disc jockeys at local radio stations who used the Internet as a source of music and international news. Several stations had made agreements with local Internet cafés, where they gave free publicity to the station in exchange for free access. Radio Lira had opened its own Internet café, which it promoted on air. Voice of Kigezi FM in Kabale had done the same thing. Journalists from Kampala-based national newspapers also used the Lira Internet cafés to e-mail stories about the ongoing war in the north back to the editors.

Across the Ugandan cases, e-mail was widely used to keep in touch with

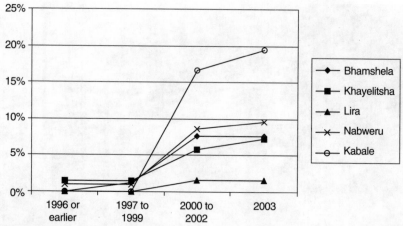

Figure 5.4 Percentage of respondents who had used the Internet, 1996–2003

family and friends. These included many people living abroad. In Lira, for example, 25 per cent of those surveyed had relatives and friends in other countries. Some people also used chat and pen-pal sites to expand their social networks internationally. E-mail was the most popular use of the Internet. Other common uses were for searching for information related to scholarships, conferences and educational programmes in other countries. Students in high school or above sometimes used the Internet for researching topics related to their course work. Entertainment was popular amongst younger patrons, who commonly used the Internet to visit the web site of their favourite football team or music group. Pornography sites were visited at almost all Internet cafés. This was unfortunate in the sense that many porn sites automatically download interfaces with shortcuts to their services, sometimes with pop-up porn advertisements. Most computers had no protection against this, so they would often become littered with self-launching pornographic pictures, creating a less-than-welcoming environment for any subsequent user who might be offended by such images. Most cafés were patronized mainly by young men, and the prevalence of pornography may have reinforced this trend.

Internet use was not yet widespread amongst professionals in any of the cases, with the possible exception of Kabale town. Those who did use it included NGO workers, professors, teachers, civil servants and health professionals looking for information related to their work. In Kabale, some people reportedly used the Internet to track financial investments in East African and international markets. The presence of public Internet services was also helpful to the tourism trade in this area, since tourists commonly sought out these facilities. In Nabweru, a few centres planned to offer distance education programmes, which would provide new educational opportunities for local residents.

Communication centres

'Those who aren't learned don't use our services – they have nothing to produce from here. Most of our users are students; they photocopy past papers for revision. We also have many working class [people]; they come to produce small documents like letters'.

Attendant, communication centre, Nansana

Communication centres are more diverse and flexible than phone shops and Internet cafés. They offer a range of services based on community demand, their own capacity, and the particular constraints and opportunities of the market. Thus, more than other type of centre, they varied with location. Their managers were often influenced by competing centres, so they might introduce a service they saw elsewhere or else be forced to stop a service if competition grew too great.

In Lira, communication centres catered to the many NGOs in the area. Some provided computer training and consultancy services to local NGOs; others offered computer maintenance. In Kabale town and Nabweru, the centres catered to other small businesses, institutions such as schools and government offices, and individuals, particularly students. Organizations would use the centres to produce and copy documents such as meeting minutes, reports, examination papers, price lists and flyers. Individuals would use them to photocopy identification papers and other important documents and for creating CVs. Some communication centres in Kabale also allowed individuals to do graphic design, edit video and copy CDs. People used these latter services for both income-generating and personal purposes. Managers of many of these centres in all three places expressed their interest in getting Internet connectivity, which might also change the mix of services they chose to offer. In Nabweru, two businesses were expecting to get distance education services – one through a partnership with a British organization offering accredited computer training, the other via a video feed of university lectures from nearby Makerere University.

In Khayelitsha, there was only one stand-alone communication centre, a Vodacom phone shop that also had other ICT-related services. Other centres offering these types of services operated within businesses or NGOs. They were used mainly for preparing CVs and funeral programmes and sometimes for homework assignments.

Computer training

'People want to have general computer knowledge because most job ads today request it'.

*Ruth Ninsiima, Computer Instructor, African
International Christian Ministry/Vocational
Training College Computer Centre*

Although many people expressed interest in learning to use computers, this did not always translate into a spending priority. Most places offering computer training had trouble attracting large numbers of students and so combined that service with other ICT-related services. Computer training was also offered at some private colleges and vocational institutes. As one service provider in Lira observed, people would not take a course until they had a definite and immediate reason to do so. People who took computer training courses were usually young and had at least completed secondary school. They took courses either because they were going on to post-secondary education or else because they felt it would increase their immediate job prospects. Often, they were people who had already had a small amount of previous exposure to computers, or who had been advised by a relative or friend that computer training would help them to find work. Women with secretarial training often took computer courses to upgrade their skills, to help them to get a job, or to improve their chances of promotion. At the Christian Ministry Vocational College in Kabale, for example, almost all computer students were women with secretarial training.

Most computer training courses offered by private centres were short-term unaccredited courses designed to help people master a particular proprietary software package such as Microsoft Word or PowerPoint, or to introduce them to the general operating system. The experience and knowledge of instructors varied greatly. Some had university degrees in computer studies while others had taken only a few two-week courses themselves. Other than testimony from previous students, most students did not have much background knowledge that they could use in making an informed decision about what courses were best.

Community service

Some communication centres were operated by people who considered the centre not simply a profit-making venture, but also a service to the community. In this spirit, they sometimes offered additional services that do not necessarily generate a profit. For example, Infornet in Lira had a plan to begin a computer club for children. Forward Communications, also in Lira, was run by a pastor who used the profits for a variety of charitable endeavours, mainly in support of orphans. This had run into difficulties because increased competition, especially in phone services, was reducing their profit margin. The centre also offered computer classes that widows were encouraged to take so as to gain saleable job skills. The fees for these classes were required in three instalments, an arrangement intended to make the training more affordable for prospective students. Some communication centres were also more willing than other types of centre to experiment with different services and technologies (often as a survival strategy in a fast-moving market), or to continue running a particular service even when it was not generating any profit, if they saw a benefit to the community. For example, a number of businesses in Khayelitsha reported running some services on a break-even basis.

Services within an existing business

These services were accessed and used in a manner similar to the communication centres. The core business, though, tended to have some influence on the types of customers who came to use the services. For example, ICT services in hotels tended to attract the hotel's out-of-town guests, as well as people living and working in the area.

Services within a not-for-profit organization

'I have been crying to be like those with access. I want to fit into the community'.

Household survey respondent, Kabale

Within schools

World Links school-based telecentres are used primarily by students and teachers, but are also open to the public. As the SBTs in Lira and Kabale were located on the town outskirts, they attracted some nearby residents who might not otherwise use the Internet. Uneducated people reportedly came only infrequently. One reason was that the schools did not always make themselves welcome to such people, who were regarded as potential thieves and vandals. One school had a policy in place banning uneducated people from the school grounds. However, the policy was waived for the telecentre which, under the conditions of the World Links programme, had to be publicly accessible. Otherwise, both schools had cheap and fast Internet connections before these were available in town. Thus, they were the first public places in both locations where people could go and use the Internet at a reasonable cost. Although dial-up had been available in both places, it was expensive and of poor quality. In Kabale, the telecentre targeted foreign tourists and also enjoyed patronage from students and professors at a nearby university. Both SBTs had concerns about how their long period offline might affect future relationships with their customers, since in both cases new Internet cafés offering more competitive services had opened up. Kabale's SBT was virtually unused by the public in the absence of the Internet connection. The Lira SBT had only a few members of the general public as users, namely those enrolled in computer training classes.

A problem common to these schools was the limited number of computers compared with the large student population. Students commonly reported frustration because their computer classes were overcrowded. Smaller, weaker students were sometimes forced off machines and these struggles frequently led to equipment breakdowns. As a result, many students did not attend computer class at all. Even those who did attend regularly – for example, the seniors in Lango College's computer club – reported that their skill levels were limited and that they needed further instruction.

Not surprisingly, the internal organization of the school influenced the success of these school telecentres or labs. Wizzy Digital Courier found that many schools in Khayelitsha actually had computers but were not using them. Schools that were effectively using the computers often had an internal champion – most effective when it was the principal or vice-principal, who encouraged or required teachers to use computers in their own work. However, there were also stories of internal politics limiting the control of resources. In Bhamshela, one school kept computers in the principal's home for safe-keeping. In another SBT, a high-level school official was reportedly trying to limit public use of the facilities because they competed with his own private communication centre in town.

Within a library

There was only one example of a library-based public Internet service in this research. Thus, it is hard to generalize. This centre, located in Khayelitsha, had limited user support, low-quality technical service, and less than ten users per day. The users tended to be well educated professionals or post-secondary students who already had experience with computers. Although the user fee of R10 per hour was subsidized by the library, it was still considered unaffordable by most local residents.

Within a community centre

The Aleksan Kopano Community Resource Centre in Alexandra was the one community centre in the study that provided ICT access. Its services included computer training, direct access to the Internet, standard office applications on computers, and photocopying. The centre also housed a community radio station.

When the centre tried to introduce a fee for Internet access, people would no longer come and use the service, even though the rate was subsidized. The centre then offered access for free and demand grew. Up to 30 people a day (the maximum the centre could accommodate) were using the services in 30-minute time slots. The library also contained some computers without any Internet connection, but with commonly used applications such as word processing. These were free but not used much. Most of the users were people already able to use the Internet. According to staff, the applications were quite broad. While entertainment was one use (but with a ban on pornography), it was not the dominant one. Some people used the Internet to seek jobs, mainly secretarial and computer-related work. High school students used it to look for bursaries and to do research for assignments. Business people obtained information on how to run a business and on budgeting and tendering, and those wanting to open bottle stores used it to seek government gazettes.[2] The computer training was intended to increase the employability of its graduates, and those attending the course were required to speak English, the language of instruction. Those attending the course were young men and women, mostly with a high school education.

Compared with private access centres, the community centre was able to offer high-quality services in a community where a commercial business might have struggled. It was also able to subsidize some services, although sustainability was always at issue and depended on strong management and willing private sponsors. Because the centre offered many community services, it also attracted many people.

Within NGOs

The types of services offered, as well as who accesses them and how, depend a great deal upon the specific NGO and its location. Two examples highlight this diversity.

The Khayelitsha Educational Resource Information Centre (KERIC) was an NGO located in a community centre complex. Its services were similar to those offered by nearby commercial centres: the centre allowed people to come and use computers, usually to prepare CVs, or else to pay one of the staff to do so at rates comparable to those charged by nearby businesses. Most users were job seekers responding to advertisements they had seen, and they would use KERIC to prepare CVs and submit them by fax or e-mail. Some students also came to KERIC to type or photocopy homework assignments. The number of users varied from day to day, but was often less than 10. The typical user profile was similar to that of people using the equivalent commercial services. The main difference was that only KERIC allowed people to use the computers themselves.

RANET, the Ugandan programme for distributing climatic information, operated out of 10 different World Vision field offices, although it was in the process of opening new sites within other NGOs. These field offices were often in remote areas and depended on their own source of electricity. They already had a computer for administration, a building, staff, a power source and a relationship with the community based on a cross-sectoral development approach. Since there was only one computer, World Vision staff would download information from the Worldspace Satellite's African Learning Channel. This included the Department of Meteorology's seasonal forecasts and other information supplied by various NGOs and research centres, including market prices supplied by Foodnet and a large range of agricultural extension information and health information. Farmers were sometimes allowed to access this information directly, but because the computer was often required for other purposes, it could not be dedicated for full public access. The information, however, was intended to be public, and with this aim it was posted and distributed to extension agents, farmers' groups and local radio stations. Some World Vision offices had also chosen to share information packages other than the climatic information (which was the explicit focus of the programme) with radios and schools. There was no charge for any of the information or training provided.

Two sites were visited in the course of this research. The Bukinda RANET site in Kabale, as with about half of the sites, ran into some early difficulties with equipment operation. In Bukinda, the person who had received training left suddenly and the remaining staff did not know how to download from the

Worldspace site. In about 2001, they sent packets of downloaded information, including National Farm Radio Forum scripts, to about 14 schools in the area and some farmers. The impact of this information was unclear, although it seemed slight. The staff received acknowledgement from the principals, nothing more. And when they stopped sending the packets, due to a computer problem, no one reacted.

The project resumed full operation about three months before the research visit. A large training effort through the Meteorology Department focused on agricultural extension offices, representatives of farmers' groups, and local representatives of the World Vision ADP management committee. The 30 participants decided to form associations that would be responsible for disseminating the climatic information to farmers within specific areas. The method of dissemination was left up to the person or persons responsible for a given area. In addition, they hoped the Voice of Kigezi radio station would broadcast the climatic information without charge.

In a RANET site in Iylowa, Tororo district, the programme was running smoothly and the farmers had had three seasons to test the climatic forecasts and related strategies. The staff had disseminated the information mainly through farmers' groups, but had not yet managed to secure the co-operation of local FM radio stations. The farmers were impressed by the results and said that so far the seasonal predictions had been accurate and helpful. While many farmers in the area traditionally depended on 'forefather knowledge' and had derided the RANET forecasts, the last few years had apparently proven RANET to be more accurate and reliable than the traditional methods. However, there was also a project under way to document the indigenous knowledge and consider ways to integrate the two sources of information and practice. The farmer groups were themselves involved in this initiative. Overall, the programme was useful and highly valued by the farmers. Dissemination remained the major challenge and limitation, and there was room to make greater use of the other kinds of downloaded information. Market price information in particular was not being disseminated.

Telecentres

'Before the telecentre, there weren't any computers. Schools got them in about 2001. So people learned about computers at the telecentre'.

Telecentre staff, Bhamshela

Universal Service Agency telecentres in Bhamshela and Khayelitsha
At the time of research, both these telecentres had extremely low use rates. Bhamshela was on the verge of closing, but was still offering computer classes to six students. At its peak, it had offered phone and photocopying services to many local residents, although other services, especially computers and Internet access, had not been used very much. It was relatively well known in

the community, although many people still knew it as 'the place that used to have phones'. Local residents were frequently interested in taking computer classes, especially for increasing their employability, but most found the price of courses unaffordable, resulting in a large stack of uncollected certificates earned by students who had never paid their fees in full.

Khayelitsha Telecentre, shown in Photo 5.2, was located in a residential area and did not appear to be as widely known. It had 10 or fewer users a day, and many of these came to use the photocopier. It offered computer classes which were attended mainly by people about to begin university who felt they needed to learn computer basics. Perhaps because these clients were only a small part of Khayelitsha's population, demand for computer training had fallen off and the classes had not been offered for about four months prior to the research visit.

Nabweru Telecentre
Nabweru Telecentre was set up beside the subcounty government offices, since these were intended to be an anchor user and the telecentre's eventual owner. By the time of the research, the subcounty had taken over ownership although it was not playing an active role. The telecentre's management was struggling to maintain activities, with little support. The telecentre's location was problematic not only because it was off transport routes and away from trading areas, but also because of its very proximity to the subcounty head-quarters. People liked to avoid this government facility because it housed both tax offices and jails – an ominous combination in the eyes of many. Perhaps for this reason, the telecentre had an average of only 11 or 12 people per

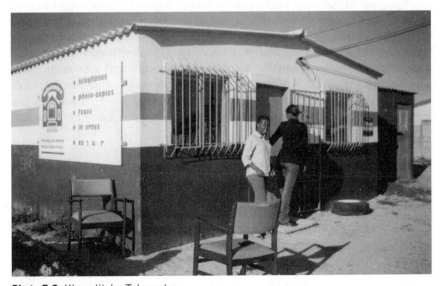

Photo 5.2 Khayelitsha Telecentre

day using its services (Etta, 2003). Subcounty staff were the telecentre's main patrons, using it mainly to drop off documents for typing or copying. The telecentre staff also engaged in some agricultural outreach activities with farmers and maintained a small library of mainly agricultural documents and videos. Their capacity to engage effectively in outreach activities was limited by lack of funds, since external funding had ended and paid services were poorly used. They were also still engaged in a number of activities through UNESCO, including an indigenous knowledge project and the installation of a community radio station.

AHI telecentres in Rubaya and Kabale

The Kabale Telecentre described itself as a 'business oriented' not-for-profit centre. At its height, it reportedly had about 100 users a day. But as competition in town increased, its customer base dropped to between 10 and 20 users a day, plus four computer students. Most of these used phone and photocopying services, while a few made use of typesetting facilities each day. Computer students came mainly from two groups of people: secondary school students on vacation and people already employed but wishing to improve their job skills. People who used the regular services generally lived nearby. Staff reflected that most people still did not use the services because they did not know about the telecentre or understand its purpose and because resources were not available for marketing or outreach activities. The centre had an Internet connection that was cut off when competition increased. When it was operational, staff reported that up to 15 farmers a week came in to access commodity related information.

The Rubaya Telecentre had about four visitors per day on regular days, and 20 on market days, once a week. It also offered computer training and had four students at the time of research. Some local schools used the telecentre to typeset and photocopy examination papers, while others continued to use stenciling as a cheaper method. The local NAADS office used the telecentre to prepare all their documents, as did local churches and subcounty staff. Community members used the telecentre's telephone but also other services for preparing wedding cards and other documents. Students and farmer groups sometimes made use of the modest library facilities and photocopied documents.

General observations

'I want to get more knowledge in using e-mail. Through e-mail and the Internet, I can get friendships outside the country. I also want to make partnerships with other organizations outside and network so we can build on our group. My training stopped because where I'm living there's no computer. I can't yet continue because all the training is centralized in Lira'.

CPAR CLC beneficiary, living in the country, outside Lira

Telecentres provide many of the same services as commercial communication centres. Sometimes they are located in remote areas where the market is not yet providing such services, as in Bhamshela in South Africa and Rubaya in Uganda. When they are in such areas, they can help extend the reach of the market by creating and/or demonstrating local demand for telecom services, but this depends in part on the availability of infrastructure. For example, other phone, fax and photocopying services opened once the Bhamshela telecentre no longer offered them; but in Rubaya, poverty, low population density and lack of electricity mitigated against the survival of ICT-related businesses.

One challenge faced by all the telecentres was low overall usage. A particular service might be in big demand, while others were not. Where commercial centres existed, they were usually unable to compete for a variety of reasons. For example, they may have been trying to subsidize non-commercial activities, or perhaps they had dial-up Internet connectivity that was unaffordable, or they were located in slightly out-of-the-way places. Because decision-making was split between telecentre management and the body overseeing the project, it was often unwieldy, sometimes tightly tied to funding conditions and stated project objectives, and less responsive to the local market, which was itself often unpredictable. Because the telecentres were generally expected to become self-financing primarily through user fees, they faced a sort of identity crisis with regard to the types of services they offered. And given that none of the centres reviewed in this research was making enough money to cover costs, they often found themselves drawn to those options that seemed most commercially viable.

Part of an integrated development strategy

'Resources aren't a problem – it's co-ordination, especially in government. There is competition between and even within government organizations'.
Karen Thorne, MediaWorks, South Africa

MPCC Programme
The Multipurpose Community Centres (MPCCs) being built by GCIS appeared to be filling a large gap in government service delivery in many communities. These were services that were required by many people, including pensioners, unemployed people, self-employed people and students. In Bhamshela, the MPCC was quite new and did not yet have a manager, but appeared to be well used. It was too early to say how the MPCC would be used and whether all services would work. It was also hard to say just how much of a role ICTs would play in service delivery. GCIS planning documents sketched a vision where staff at the different offices housed in the MPCC would play the role of information intermediaries, helping to answer people's questions and referring them to different information links and sources, including ICT-related ones such as phone numbers and Web pages (GCIS, 2001). Whether this would

work in practice would depend greatly on the staff themselves, the types of questions asked by community members, and having reasonably priced and reliable ICT services either at the MPCC or nearby. In 2003, the high cost of Internet access and lack of connectivity options other than dial-up meant that most MPCCs did not have an Internet connection and were unlikely to get one until something changed in the national telecom market.

NAADS Programme

The National Agricultural Advisory Service (NAADS) is a large 25-year programme begun in 2001 by the government of Uganda to 'increase farmer access to information, knowledge and technology for profitable agricultural production' (NAADS, 2001). It is part of the government's Plan to Modernize Agriculture (PMA). The first seven-year phase costs an estimated US$108 million, funded primarily by outside donor agencies. By the end of this phase, the programme is expected to have a presence in every district of the country.

NAADS replaces the old government extension system with a new decentralized demand-driven extension system within the private sector. The programme gives farmer groups resources with which they can choose to contract extension operators to provide them with certain information and inputs with the ultimate aim of commercializing farming in the country. Once they have adopted practices and inputs that allow them to generate a profit, they should, in theory, be able to pay for extension services themselves, thus rendering those services sustainable.

An early criticism of NAADS from a variety of observers has been that, despite the explicit aim to target poor subsistence farmers, those without resources are unable to participate in or benefit from the programme (eg, FOWODE/Oxfam, 2004).

From the perspective of universal access, NAADS provides interesting potential in rural areas, since creating demand for information is a major obstacle faced by telecentres. If NAADS succeeds in creating a network of farmer groups who have the capacity to demand information, rural telecentres may serve an important intermediary role. In Rubaya, the AHI telecentre was already collaborating with NAADS, and the staff there had helped it to identify and expand the number of farmer groups in the area.

Radio and Internet

In South Africa and Uganda, as elsewhere in the world, radio is the most far-reaching method of broadcasting information. The government of Uganda found that in the mid-90s the state-run Radio Uganda was the main information source for Uganda's mainly rural population (Government of Uganda, 2002). In both countries, the liberalization of the airwaves is a relatively recent phenomenon, starting in the early to mid-90s, the same period in which the telecom sector began to be privatized.

In Uganda, private operators could apply for FM licenses. By 1998 there were 28 FM stations in the country and by September 2003 this had risen to 119.

One reason for the quick spread of these stations was that station owners were making a profit, mainly by signing up big advertisers and by receiving sponsorship from government and other agencies to broadcast development-related programmes. In Lira, three radio stations were located within the town (Radio Lira, Radio Wa and Radio Rhino) and had all opened up within the past three years. All three used the Internet to obtain material for programming. Radio Rhino had an agreement with a local Internet café whereby they were able to browse for free in exchange for promoting the café on air. The same scenario was repeated in other towns across the country. In Kabale, the largest radio station had also set up the only financially successful Internet café. RANET was also pursuing relationships with stations in its areas of operation to have climatic information broadcast so that it could reach a wide audience. About six sites, some radio stations and some telecentres, were being converted to 'community multimedia centres' under a new UNESCO initiative. This set up community radio stations within telecentres (or conversely, telecentres within radio stations) in an attempt to maximize the benefits of such a symbiotic combination. At the time of research, the installation was in progress.

In South Africa, radio was more tightly regulated and there were fewer private stations. But the community radio sector was very strong. Ninety-six community radio stations had been granted licenses by the end of 2003, and over 100 (some still awaiting licensing hearings) were members of the not-for-profit National Community Radio Forum (NCRF). The Forum lobbied on behalf of community radio at the policy level, helped stations with programming, especially around events such as national elections, and assisted with capacity building, specifically the incorporation of new ICTs into radio operations. A major focus was satellite communications for live-to-air broadcasting. Its longer-term goal was to introduce two-way satellite connections so that stations could directly share programming with each other. It was supporting stations through 'hub' stations in each province which shared their equipment and expertise with neighbouring stations. Unlike Uganda, there had been little direct interaction between community radio and telecentres, and what little there was reportedly did not have much success. The MPCC programme had planned that its facilities would house both telecentres and community radio, though it was unclear what relationship might develop between the two.

The two countries have distinct experiences in terms of radio and its interactions with new ICTs. However, both demonstrate that radio remains the most popular way to receive information, since it is low-cost and widely accessible, does not require literacy or special skills of its users, and is usually available in local languages. Telephone and even Internet use (which is dominated by e-mail) tend to be used for personal communication purposes most of the time, rather than for seeking out new information. Combining new ICTs with radio clearly increases the value of the information they can provide by greatly increasing its reach.

Analysis by livelihood

Survival

'I want to know how to use a computer because people are growing heavier with insecurity here. I don't know how, but maybe it will help'.

Household survey respondent, Lira

People in survival situations usually do not have the resources to use ICTs even when they are available. Depending upon the situation, some people may have a reason and the capacity to make a phone call, usually to a family member. People in Lira would sometimes go to the radio stations in times of crisis, to find accurate up-to-date news and to try and track down family members.

The greatest impact that ICTs may have on people in survival situations is indirect. NGOs, government departments and aid agencies that serve populations in urgent need may better co-ordinate their efforts and get useful information if they have access to telephones, computers and the Internet. Another major indirect impact of ICTs on such people relates to the smoother flow of information to broadcast media and the availability of more diverse and timely information sources, especially in situations where a dominant group may try to limit certain forms of information. In Lira, for example, radio stations depended upon village reporters for news. Mobile phones, where the network was available, could help these people to transmit information. E-mail helped journalists from Kampala to submit stories quickly for publication. These uses of ICTs can help to build public awareness of emergency situations. The conflict in northern Uganda is a good example. It was often not treated as a major issue amongst people in the south. Nonetheless, and in part because of good communications links between Lira and the rest of the country, national daily papers often covered the situation, thus helping to keep the issue fresh. These reports contrasted with the official government news releases which tended to downplay the level of the conflict and announce that the end was imminent. Sadly, whatever impact honest media coverage may have had on national public opinion and government commitment, it had not been sufficient to change the situation.

Another potential benefit of promoting ICTs is suggested by an anecdote from Gordon Bell, owner of Radio Lira. He recalled an incident in which a person had died of a treatable condition because a local doctor had refused to give the cure – which itself was not expensive – because the patient could not pay a hefty consultancy fee. To Bell, this demonstrated that information, when limited, was held by the powerful at the expense of the weak. If people had a resource like the Internet at hand and know how to use it, they could look up basic medical information that might literally save a life.

Subsistence and subsistence plus

'I don't need to learn how to use computers because I am always out working in the field'.

Household survey respondent, Rubaya, Kabale

'Climate patterns are changing – seasons and rainfall. So farmers need advice about which crops to plant and when. If the information is too slow [in reaching them], it is useless because it doesn't influence the decision making of the farmer'.

Milton Waiswa, RANET

For people in primarily subsistence lifestyles, the radio is a widely used source of information. Those who use the telephone do so primarily to maintain contact with family members in other parts of the country or other countries. These contacts are often used to request money or help acquiring needed items. Where the Internet is available, e-mail, and to a lesser degree, chat, is also used for maintaining international contact with the Ugandan Diaspora.

Although there are many situations in which information is useful to specific livelihood strategies, whether related to agriculture, markets, health or education, people in this category do not tend to actively seek out new information. They are more likely to depend on traditional forms of knowledge already embedded in their practices. As there are already plenty of demands on the little money they have, their willingness to pay for information services is usually very limited. For example, the potato farmer in Photo 5.3 has hired help in harvesting her crop, but says she cannot afford to use ICTs and has no real reason to use them. Even when it is free, information is not likely to be applied unless it actively builds on existing knowledge or creates new social structures. For example, RANET Uganda provided farmers with training about what to do in different climatic conditions. Where this conflicted with traditional local knowledge, more traditionally minded people watched to see how those who adopted the new practices faired.

On the one hand, then, telecentres have had a difficult time reaching these people, even when they are physically accessible, because user fees for services generally outstrip ability and willingness to pay. The exception is phone services. On the other hand, these populations may benefit indirectly from telecentres if the local institutions which serve them are able to operate more efficiently as a result of being able to tap into locally available ICT services.

Self-employment

'I want to know more about repairing computers. There are a lot of benefits to that kind of knowledge now. For example, most jobs at the NGOs require

Photo 5.3 Potato farmer in Rubaya

you to be computer literate. This knowledge could also help me to open my own training school if I can get loans to buy computers. Then I'd be self-employed'.

Participant in World Bank-sponsored Youth ICT
Programme, Lango College, Lira

For self-employed people, both phone shops and communication centres provide useful business support services. Phones can be used to contact suppliers, clients and potential clients. However, personal cell phones are also very popular and provide a useful feature that communication centres do not offer: they make it possible for business contacts and clients to call at any time. For example, some taxi drivers in South Africa and Uganda carried cellular phones, as did some traditional healers and people working in many other types of service and trading.

On the productive side, providing ICT access and value-added services is an opportunity to make money that many entrepreneurs in Uganda, and to a much lesser degree, in South Africa, are seizing. However, their early experiences show that these ventures often have high risks and low profits. Part of the reason for the low profits is that, even where demand exists for phone and Internet services, the high rates charged by telecom operators, and to a lesser extent ISPs, do not allow much room for mark-up. However, these entrepreneurs are crucial players in the pursuit of universal access since it is through their centres that the majority of users gain access to ICTs. They also assist large telecom operators by helping to open up telecom services to a larger segment of the market. Their risk and added value should be somehow recognized and they should share in a greater proportion of the financial benefits from telecom sector growth. This would also ensure that more money circulates locally instead of being sucked out of the community (and often the country) by the major operators. The positive example of how such a model can work is the Vodacom phone shops.

Formal employment

'I'm studying computers because it will help with job opportunities. After I'm finished, I want to work at a computer till'.
Computer student, Bhamshela telecentre

Computer skills are strongly associated with formal education and formal employment. If all people are to have the opportunity to learn computer-related job skills, computer training should be available in schools. But because many people already struggle to gain access to formal education, especially at secondary and post-secondary levels, this creates the potential for a greater social and economic divide within a country. Some schools do not have the resources to acquire computers and many do not have electricity. In any case, the presence of computers does not by itself guarantee any direct benefits. Often there are few computers available, a lack of qualified teachers, and problems associated with security, infrastructure and cost. Many schools pay for computers through increased school fees, and sometimes through a special computer lab fee. The poorest rural schools in Uganda, even if they managed to get computers, would not be able to impose these fees on the students as most families would not be able to pay. Short-term private classes are available in many areas, but the fees are often prohibitive and standards vary widely. This is a major policy issue at the national level.

This research did not quantify the benefits of knowing about computers or identify exactly what computer skills were in demand, how many skilled people managed to find jobs, or where they found them.[3] For most people in the case communities, the other assets required to benefit from a computer-skills strategy, plus the investment of time and money, were too great to even consider. However, many young people were interested in gaining computer skills to improve their future job prospects. Certainly, some people with computer skills, mainly men, had chosen to start their own businesses or work as consultants, either in Kampala or in smaller towns. Women tended to gain skills for work in secretarial and office support positions.

Without advocacy for the less powerful local businesses, large international companies are likely to take the lion's share of profit in the ICT sector. Like-wise, without equal access to quality education, the divide is likely to increase between those who can afford well-equipped private schools and courses and those who cannot. Both these issues require national-level intervention and are arguably some of the greatest problems related to universal access.

Connections within the extended family are one of the greatest equalizing forces in this otherwise highly-stratified scenario. Those able to go on to higher education and formal employment often end up supporting members of the extended family and even friends. In this respect, it is common for poor families to support one member to pursue higher education, thus increasing his or her job prospects.

Acquiring assets

'Most women said they don't need information, they need credit . . . we need to move beyond the tangibles'.

Florence Kuteesa, CEEWA

'I'd like to know about computers, just to have some knowledge about it at first. We the women who have children to look after need some of these services, such as computer knowledge, which may be a source of income'.

Household survey respondent, Lira

The ability to use ICTs often depends on possession of other assets – money, mobility to get to an access point, literacy or, in the case of Internet use, the ability to understand English or another major world language. Access to, and ownership of, ICTs are valued as assets in several respects:

- They are status symbols. Examples included owning a recent-model cellular phone, being able to use a computer, and having an e-mail address.

- ICT-related skills, and especially computer skills, can improve one's chances of getting a good job.
- ICTs, especially telephone and e-mail, can be used to maintain contact with family and friends across the country and the world.
- The Internet can help people create new international links with others. Young people particularly value this attribute and sometimes see it as a potential source of economic gain. For example, a boy in Lira reportedly made contact with someone via the Internet who helped him pay his school fees.
- For many people, mastering basic ICT skills boosts their own self-confidence, since ICTs are popularly associated with the modern world. As one woman in Bhamshela commented, if she and her family cannot use computers, people will think they are uncivilized.

However, information per se is not highly valued, although many people listen to the radio and value its benefits. A few organizations that have worked with community groups commented that people did not view information as a resource of major importance. CEEWA Uganda, for example, noted this in work with women entrepreneurs, and AHI in its dealings with farmers in Rubaya.

Maintaining and expanding family and social networks

'These phones are important because they allow us to communicate with our relatives'.

Phone shop user, Bhamshela, South Africa

As already noted, communication with family is the single largest use of the telephone and, in the case of people with family members abroad, it is also a major driver of e-mail use. Such communication is important because of the remittance economy: people employed in cities often send rural relatives money and other requested items, and sometimes host young relatives seeking work in the city. ICTs help maintain these networks. Also, when people with ICT skills use them to gain entry to the formal economy and better paying jobs, they often end up sharing the economic benefits with rural family members. This kind of use may act as an equalizing force within a system that otherwise favours those with higher-level formal education and jobs in the formal sector.

Impact and strategies for maximizing benefits

From a livelihoods perspective, what are the major positive impacts of ICT access on development? Here are four key benefits, the first of which is the most important:

- ICTs support the remittance economy and help maintain contact between geographically dispersed family members.
- They create more opportunities for interaction with other parts of the world, and these may, in turn, lead to new economic opportunities.
- They support independent media coverage through e-mail and online news.
- They create new opportunities for small business. However, some of these ventures, especially those involving computers and the Internet, pose high risks and generate only modest profits.

ICTs can also have a negative impact on livelihoods. Using them is expensive and draws on scarce financial resources. Young people in particular may spend a large proportion of their income on cellular phones and access time at Internet cafés. While such expenditures may at times constitute a form of investment, the returns are unlikely to equal output in economic terms. Indeed, ICT usage is often motivated by a desire for higher social status or entertainment. Most of the profit generated from using ICTs, whether through access centres or private ownership, goes to big telecom operators, which are in a much better position to lobby for their own interests than are small local businesses. Finally, the growing demand for computer skills for certain types of jobs and for university studies (more so in South Africa) creates higher entry barriers to both post-secondary education and formal employment. This, in turn, widens the chasm between poor and rich families, and especially between rural and urban families.

Based on the experiences of the access centres reviewed in this research, there are a variety of strategies that can be used by the centres themselves and by policy makers to maximize the positive development potential of ICTs and reduce some of the risks.

Access centres

A basic distinction can be made between phone and other ICT services, in terms of accessibility. People appreciate the value of telephone services without any great need for awareness building, special skills or education. Physical availability and cost are the two factors determining whether and how often a person uses the telephone. Access centres are cheaper than private ownership although they do not of course offer all the same advantages: most will not accept incoming calls and many also lack privacy.

For ICTs focused on information acquisition, development impact is maximized where they are integrated into other activities and structures. Access focused clearly on one application can build capacity and lead to further exploration of the possibilities of both the technology and the information available through it. RANET is an excellent example of this.

For computers and the Internet, there are multiple barriers to use. For many populations, direct access is not feasible; mediated access is more likely to bring benefits. Most access centres have had only limited success in addressing the barriers their potential customers face and only certain populations used their services, particularly professionals and students. In many cases, those using access centres had some exposure to computers in another setting such as school or work.

User fees alone are not necessarily an insurmountable barrier to use; but combined with other barriers, they create a major disincentive. For example, most people within a 20-minute walk of phone service are willing and able to pay. These rates are often high by international standards. South Africa, for example, has some of the highest landline phone charges in the world. But poor people value and use these lines – and they control costs by limiting the length and frequency of calls.[4] In the case of computer and Internet services, three factors combine to minimize use: user fees, lack of knowledge of potential applications, and inability to actually use the technologies. Inexperienced people do not explore the potential of the services since there is no immediate benefit in doing so. Using computers and the Internet in a way that has tangible value takes some investment in terms of time; and when time also involves money, this increases the cost of the investment. Youth are the people most likely to use new ICTs at access centres without prior exposure. They are motivated purely by curiosity. But for most adults trying to get by on few resources, curiosity with a price tag is not worth indulging. This implies that fee-based access to computers and the Internet is unlikely to create widespread popular use, unless there are other places where people can build proficiency in these technologies free of charge.

The effectiveness of any access centre depends on its internal capacity. Dedicated, energetic leadership and management, along with the ability to partner effectively, are particularly important characteristics. In addition, the technology itself must be reliable, or at least the organization using it must be resourceful – meaning that if the technology does fail, there should be an alternative path forward while the problem is being addressed.

Access centres can also have a positive impact on development by increasing local job opportunities, building staff capacity in ICTs and management, and participating in other community development initiatives. These contingent benefits should be explicitly recognized and targeted where possible. Externally funded projects in particular need to maximize local input and control from the beginning – both to build capacity and ensure local relevance.

Policy makers

If one compares common patterns of ICT access and use with policy visions and the great promise of ICTs, it is apparent that such potential cannot be realized spontaneously through the market. One factor that limits the use of ICTs as components of livelihood strategies is the common perception that these technologies, especially computers, are for educated people only and primarily needed for enhancing job opportunities. Such a popular view limits the spontaneous appropriation of ICTs where physical access is provided. This underscores the earlier argument that ICT services need to be harmonized with larger national development strategies, especially those for rural areas.

Another clear issue is affordability. Both end consumers and local access centres, whether run by entrepreneurs or NGOs, struggle with the high cost of equipment and telecom services. These costs can limit the spread of ICTs and undermine their potential value. Affordability itself is a very elastic term, since it is often difficult to place a clear value on an ICT service – it depends on what it is used for. The International Telecommunication Union, for example, suggests 3 per cent of income should be spent on ICT-related services; but in Khayelitsha, based on people's own reported spending estimates and census figures, people were spending over 25 per cent of their incomes just on phone services! What then, is an affordable rate?

Larger urban centres have better infrastructure, a wider range of services, and greater demand. However, services tend to target the wealthier segments of society, allowing for easier profits. This tendency is especially evident in areas with high levels of inequality. In South Africa, most urban cybercafés cater to the middle classes; but in the townships there are, with the exception of phone shops, very few services.

Even small towns in Uganda can sustain some ICT services, aided greatly by local Internet points-of-presence. Again, the usage patterns indicate that the market tends to focus on a small segment of the population – professionals and other well-educated people. In rural areas, however, there are few wealthy people and very few ICT services. In these areas, the market generally cannot sustain services, with the exception of some personal cellular phone services that are often informally shared and resold.

These distinctions are sometimes blurred, especially by outside agencies, in identifying areas and methods for supporting universal access. Small towns are sometimes identified as 'rural' and the degree of existing economic activity is not recognized. Provision of public access can be seen as a valuable end in itself, whereas in practice it duplicates and competes with services already available. Access initiatives in such locations may still be valuable given that overall use and awareness are low, but they need to be much more purposive and targeted if they are to reach a different segment of the population and meet a felt need. For example, rural farmers may not know what ICTs can do, but they may express a need for timely market information or weather forecasts.

Co-ordination and communication across government sectors and levels are important if ICT-related infrastructure development efforts are to be properly linked with intended uses. Both Uganda and South Africa have engaged in this to some degree, and their experiences indicate that the process of change is lengthy and profound. In South Africa, the GCIS is facilitating a multi-stakeholder process to design and deliver MPCCs, whereas in Uganda NAADS is co-operating with existing telecentres and other ICT projects and can help to build effective demand for these services. This type of co-ordination can also help to prioritize and identify appropriate technologies. For example, RANET is using Worldspace data satellite receivers which allow its sites to receive a large amount of data, but do not allow for uploading. While this is a more limited technology than the Internet, it is much cheaper. Such co-ordination can also help avoid duplication. For example, before GCIS began its co-ordinating work, a number of government departments were implementing or planning similar community centres without any clear links.

Multi-sectoral links also make it possible to constantly revisit the goal of universal access in the context of current national and local development priorities. Where universal service funds exist, it is possible to target access services in ways that maximize their reach and developmental impact. For example, World Vision wanted to step up their RANET sites into public access sites with a dedicated computer. If they could propose such a venture and obtain a one-time grant from the UCC to acquire the initial equipment, similar agencies might follow suit.

CHAPTER SIX
Conclusions, lessons and research directions

South Africa and Uganda in the broader context

The conclusions, lessons, and suggestions in this final chapter are based largely on the empirical evidence from the two countries covered by the study, Uganda and South Africa. How widely applicable, then, are the lessons that can be drawn?

South Africa and Uganda are two specific examples of a process that is occurring, with different variations, across Africa and the world. Firstly, national governments and international agencies have since the 1990s, paid increasing attention to the importance of information and communication technologies, since these are drivers of the new emerging 'knowledge economy'. Secondly, 105 national governments made commitments to the WTO to privatize and liberalize their telecommunication markets under the GATS agreement beginning in 1997 (Bressie, Kende and Williams, 2004). Uganda and South Africa were two such governments. Since these agreements committed governments to similar conditions and processes of telecommunications reform, it is not surprising that some early results reveal similarities across countries. This reform has also influenced the penetration of newer telecom technologies such as the cellular phone. In this case, there were generally multiple licensed private service providers early on, in contrast to the situation for fixed-line telephone provision, which in most countries was rooted in public monopolies. The Research ICT Africa network has produced some interesting cross-Africa comparisons of telecom sectors in seven African countries (South Africa, Uganda, Cameroon, Ethiopia, Rwanda, Kenya and Zambia). These comparisons show that, while different countries have taken different routes to meet their commitments, common patterns exist. Most countries encounter barriers to achieving 'ideal' liberalization. Weak regulators and political lobbying by powerful private and state interests, for example, are fairly common. However, those countries in which the telecom sector has been more fully opened up generally appear to have better overall access at lower prices. Uganda, although far from being in a state of perfect competition, has gone further than most African countries in liberalizing its markets. In contrast, Ethiopia still has a state-run monopoly and tele-communication services are scarce, expensive and difficult to obtain (Adam, 2004). South Africa is in many ways a unique case, due to its higher GDP, much stronger existing telecom infrastructure, and high levels of inequality.

In short, applying lessons from South Africa and Uganda requires an understanding of the contexts in which the observations were made and in which they are to be applied. Referring back to Figures 1.1 and 1.2, we can reconsider Uganda and South Africa as two specific instances of a general pattern occurring within Africa and beyond.

Examining key differences in ICT prevalence and use between South Africa and Uganda also points to differences in causal factors. Some of these are noted in Figures 1.1 and 1.2, while other aspects, especially those related to local culture, have not been fully captured and explored by this analysis. This is true, for example, for the greater gender imbalance amongst ICT users in Uganda than in South Africa. Table 6.1 shows some of these differences in causal factors and resulting ICT accessibility, use or developmental impact.

Table 6.1 Differences in ICT accessibility and use in South Africa and Uganda

Differences in causal factors	Results (ICT accessibility, use, developmental impact)
Policy intent	
Greater national focus/activity by government and NGOs in the community media sector in South Africa: community radio, posters, newspapers.	A substantial role for community media within the MPCC model in South Africa, with no equivalent in Uganda.
Major focus in South Africa on government service delivery, redressing historical inequities.	Greater implementation focus in South Africa on ICTs to deliver government services.
Implementation	
More of a direct implementation approach in South Africa, compared with regulation/subsidy and focus on market means by Uganda.	An underperforming telecentre programme in South Africa which is too limited to reach most people. In Uganda a wider variety of approaches based more typically within the market, but with large questions remaining about accessibility to rural populations.
Larger direct donor involvement in Uganda, and greater interaction between donors and the Ugandan government.	A wider range of telecentre models in Uganda compared with the mostly homogenous Universal Service Agency telecentre programme in South Africa. Early donor experiments influenced Ugandan policy makers' decisions about how to proceed on universal access.

Differences in causal factors	Results (ICT accessibility, use, developmental impact)
Less effective regulation in South Africa, and earlier competition (especially in fixed line telecommunications) in Uganda.	Fewer available technological options (eg, limited availability/deployment of VSAT, wireless), especially in rural areas (urban areas in the process of getting ADSL) in South Africa.
National/local context	
In South Africa, a strong history of resistance to an oppressive state, with a lingering sense of distrust and disenfranchisement ten years after the end of apartheid.	High rates of theft and vandalism, even in 'community- owned' projects.
Artificial, economically weak settlements (ie, townships) in South Africa versus functional towns in Uganda.	Less overall economic activity within townships makes beginning a viable computer-based enterprise more difficult, although phone services have ready demand. Ugandan towns, in contrast, appear able to sustain a much wider variety of services, including training, computer secretarial and desktop publishing services, as well as Internet cafés. Phone services, however, remain most prevalent and appear most likely to generate profit.
Different migration patterns: in Uganda a greater international Diaspora; in South Africa, almost exclusively internal migration in the communities studied.	Greater apparent demand for e-mail in Uganda and high demand in both countries for telephones.
Varying local economic and cultural factors.	Gender gap: fewer women amongst ICT users in Uganda compared with South Africa.

Similarities between South Africa and Uganda in ICT access and use can equally be explained by corresponding similarities in policy intent, political and economic situations, and external factors. As already mentioned, the overall processes in each country were shaped by the WTO's GATS trade agreement. Both countries had relatively high costs of connectivity, related to still limited markets with powerful telecom operators which could be difficult to regulate. Equally, those managing access centres in both countries related stories of unresponsive service and incorrect billing from landline providers. In both countries, local political troubles had sometimes turned a subsidized access centre into a source of conflict, and some observers had concluded this was another argument for privately owned centres.

Lessons for policy makers

1. Market liberalization is more likely to lead to broader access than telecommunication monopolies with some obligations – especially where the regulator is weak.

Based on South Africa's experience, a monopoly telecommunication market results, not surprisingly, in high costs and a lack of response to consumer needs. The argument often made for telecommunication monopolies – that they are necessary for achieving universal access – does not bear up in the experience of either South Africa or Uganda. In both countries, the relatively more open and competitive cellular markets were responsible for the majority of new telephone users. This increase was never a major part of the original universal access strategies since the popularity of cellular services was not fore-seen. Although this is generally a positive development, cellular phone use is still much more expensive than landline phone use, and cellular networks are not well-suited to large data transfers. While new technological developments may change these limitations, some close observers have argued that public investment in telecom infrastructure, including fibre optic cable, is necessary to move a nation towards universal access and to maximize the potential benefits of the information economy (Tusubira *et al.*, 2003).

For Internet use too, the relatively more liberalized environment of Uganda has facilitated universal access. Although South Africa has the largest propor-tion of Internet users in Africa, it is also one of the most unequal countries in the world and patterns of access reflect this. Rates of Internet use in Bhamshela and Khayelitsha were lower than in many places in Uganda. In both places, dial-up had proven a poor option for Internet connectivity because it was expensive and of poor quality. Only in Uganda (and urban areas in South Africa not covered by this research) had people been able to use alternative methods of connectivity, including VSAT and wireless broadband.

2. Universal access and access centres need to be integrated into national development strategies, especially in rural areas.

Both South Africa and Uganda demonstrate that, left to market forces, certain populations are likely to be excluded from certain types of services and bene-fits associated with ICTs. Moreover, the ability to benefit from ICTs depends upon the provision of services that have traditionally been within the domain of public services – especially electricity, roads and education. At the same time, ICTs may play a complementary role in larger development strategies. In South Africa, the Multipurpose Community Centre programme focuses on creating structures for government service provision and community media in traditionally disadvantaged areas. In Uganda, the National Agricultural Advisory Service, which is integrated into the overarching national develop-ment strategy, focuses on creating demand and linking it with the supply of agricultural information. Both of these programmes can benefit from the presence of basic ICT infrastructure in rural and disadvantaged areas, and can in turn act as 'anchor clients' in creating effective demand for these services, thus contributing to their financial viability in poor areas.

In 2003, both South Africa's MPCC programme and Uganda's NAADS strategy had been operating for a few years, but it was too early to determine what the long-term viability and impact of these strategies would be. Both required the creation of new organizational structures and new working relationships, within government and between government and other stakeholders. This was a time-consuming process that committed to a longer-term vision. Such types of activities can be viewed as social investments in underdeveloped communities. If carefully co-ordinated and adequately resourced, they will create opportunities to ensure ICTs serve the public good.

Any attempt to integrate ICT access into development strategies will be strongly affected by the larger market and regulatory environment. Uganda, despite the fact that the national ICT policy had not yet been passed, seems to have faired better than South Africa, which had, from the beginning, an explicit policy and a special agency to focus on universal access. This was in part because much of this policy, while ambitious, was not clearly defined and planned out. Also, the South African regulator was generally perceived to be weak, especially because of regular interference from the Ministry of Communication. Any integrated ICT policy will be seriously hobbled unless it includes clear definitions and action plans, and operates in an environment where the national telecom sector is effectively regulated.

South Africa has experimented with direct government intervention in universal access service provision. As of 2003, these efforts had not been very successful. This cannot be said to prove that direct government intervention is always a bad idea – but rather that it is affected by larger market and infrastructure issues. The Universal Service Agency might have been more effective had it been set up as a branch of the regulator. As such, it would enjoy the same independence and power vis-à-vis the operators as the regulator, and universal access issues would always be framed within the larger context of the telecommunication market.

Besides being highly dependent on the wider national market, access centres operating through user fees are not adequate by themselves to build local demand. This is because of the multiple barriers to access that many populations face. Low levels of education, illiteracy, low awareness, lack of access to transportation, and lack of money all contribute to low levels of ICT use. They are reinforced by a popular conception of ICTs, and especially computers and the Internet, as the tools of the educated only. Most people are concerned with gaining immediate productive assets. Knowledge of computers is usually seen as a long-term investment for people who want to seek formal, professional jobs opportunities. While this understanding reflects current reality, it also overlooks the potential that these tools hold for other types of livelihood, and especially the potential value of information in productive activities.

Unfortunately, this perception is widespread and such uses are therefore unlikely to be discovered spontaneously.

Because knowledge of computers is necessary in a growing proportion of formal-sector jobs, public schools are also important potential equalizers amongst the population. Creating a strategy aimed at giving all public school students a fair chance to use computers and learn related skills is therefore another important complement to universal access strategies. However, there are huge obstacles to achieving this goal in both South Africa and Uganda because many schools are under-resourced and have no electricity. Also, the presence of computers in schools by no means ensures that all students will receive effective instruction in their use or get the chance to apply them in other areas of the curriculum.

3. **Large telecom operators seem to profit disproportionately from the telecom services offered through shared access centres, with little of the profit staying in local communities. Regulation can improve this situation by recognizing the value of the service provided by local intermediaries between big operators and the public.**

Large telecom operators have much greater political clout than the fragmented groups of consumers and small businesses further downstream. Access centres in local communities are often small local businesses or not-for-profit organizations that actually help generate profit for the large operators by aggregating demand and creating use amongst people who would not be able to use the services by other means such as private ownership. In South Africa especially, the gap between rich and poor is one of the largest in the world, and it is therefore tempting for the large operators to ignore poorer markets. Again, it is the small entrepreneurs who often help to bridge this gap. Regulators can do much more to explicitly recognize the valuable contributions of these service providers both to the public and to the large operators. The latter do business in monopoly or near-monopoly situations, whilst the former face fierce local competition. For these reasons, regulators should seriously consider requiring large operators to give preferential ('wholesale') rates to any agent who either resells public access to telecom services or provides them free.

4. **There is a clear practical distinction between phone and Internet access: the former has much greater local market demand than the latter.**

The distinction between phone and Internet access may reflect a broader distinction between communication-related and information-related services, where the former commonly enjoy greater demand than the latter (Odlyzko, 2004). In terms of policy, it implies that it is worth thinking of telephone services as separate from, or at least as not necessarily connected with, the Internet. This is especially true as long as GSM (Global System for Mobile Communications) plays a major role in providing basic phone access, since the same technology is not suited to large-scale data transfer as required by most

Internet applications. However, where and when VoIP is legalized, or GSM-related technologies become better suited for broadband data transmission, this distinction might once again cease to be important in terms of physical infrastructure. It will remain relevant, however, in terms of barriers to use. Telephones are simple to use and do not require literacy or knowledge of a major world language. Most computer-based applications, in contrast, require literacy and some training and time for mastery. And they are vastly more useful when the user has knowledge of a major world language. Because of this, any strategy to provide universal access to computers will need to be closely linked to public education and adult education programmes.

Lessons for those implementing access centres

1. Set a clear and simple initial goal that is based on a real local need and on which the centre can build.

Early telecentres were often conceived as extremely ambitious projects, offering a full range of services from virtual learning libraries to telehealth applications. But in South Africa and Uganda, they were largely unable to deliver on these high expectations. One of the biggest problems seems to be with the expectations themselves. For example, local stakeholders, including telecentre staff and community members, were often introduced to telecentres with great tales of what they could bring. Whatever the centre did manage, it almost always fell far short of these expectations. Local support for telecentres, essential to their continued survival, often decreased over time and staff became demoralized. Telecentres can reduce risk of big disappointment and lowered credibility by starting with modest expectations that they can in fact fulfil. This also allows local capacity to be built gradually. RANET, for example, began by focusing on the delivery of seasonal forecasts. As many of its sites flourish, it is natural for local stakeholders to experiment with other applications of the informational resources available to them. Any intervention by outside agencies also requires an honest assessment of risk and potential limitations.

2. Externally subsidized access strategies should be based on a full local market assessment that takes into account both current conditions and likely future developments.

Another hard-earned lesson from the early telecentres is that outside agents often did not assess the local market. Rather, they tended to assume that ICT access services were in short supply and that they should therefore go ahead and set up a centre to meet a need. This assumption was often true when the original centres were launched, but the local market often changed quickly – and not always in response to the telecentre itself. It also links to the sometimes arbitrary distinction often made between 'urban' and 'rural'. In the implementation of funded telecommunication projects in Uganda, 'rural' was sometimes used to refer to any place other than the capital city, Kampala.

While Kampala is the largest, most important urban centre, Uganda also has large towns. 'Rural' telecentres situated on the outskirts of these towns are still competing with a small but active town market. In Uganda, many of the larger towns have hosted some kind of access centre since about 1998 – the same period in which the first telecentres began.

This implies that those implementing funded access centres should consider how these are likely to support local efforts and meet local needs, rather than usurping and undermining existing local efforts. Perhaps they are providing a particular service that the market cannot, or they are providing access for a particular segment of the community that market-based access centres cannot reach. They should further consider how they might work with and perhaps directly support local efforts, whether they are small businesses or NGOs. Supporting existing efforts is likely to be more cost-effective and to have greater development impact than unilaterally creating a new institution. It may add value to the work of the existing organization. Many local entrepreneurs also have an interest in their community that goes beyond a pure profit motive. Supporting them will in turn support the local economy and may create jobs.

3. **Control for complexity and localize decision making as much as possible. Funding needs to be flexible and responsive to local market changes.**

Local markets can change rapidly, especially with regard to ICTs. So access centres need to be able to respond quickly and flexibly. This requires a lot of locally based decision-making power, as well as capacity to make decisions. Good management has long been recognized as a key to telecentre success (eg, Fuchs, 1997). In both South Africa and Uganda, early experience also showed that placing scarce resources, in the form of telecentres, in resource-poor communities could lead to local political conflict, especially where the centre was under the control of local government authorities. This is another reason why entrepreneurial models sometimes seem to work better. For some of the people involved in the early Ugandan telecentres, this was the single most important lesson from their experiences. Certainly, a clear, simple decision-making and accountability structure is important. Telecentres that rely on boards filled with local politicians and funding decisions tied to a preplanned budget or a centralized authority are likely to run into serious operational difficulties, as these structures will not allow them to respond to community needs.

4. **It is generally not realistic to expect a centre to be profitable immediately after start-up.**

Financial viability over the short term is most likely where:

- there is already demonstrated need and willingness to pay;
- a centre is able to count on service of fair quality and reliability;

- running costs are not high and profit margins are fair;
- the access centre is either very simple, or is an addition to an already functioning organization;
- the access centre is in a highly visible place where people come regularly to shop and do business.

User fees will often play a role in sustaining access. This may not be a bad thing, insofar as willingness to pay is an indication of local relevance. However, the barriers that many people face in gaining access to ICTs and using them, especially computers and the Internet, are multiple and cumulative. User fees are an extra impediment for a large segment of the population. Thus, for any particular centre, depending on its aims and intended reach, it is important to consider the likely impact of user fees and the pricing structure on the services offered. For telephone services, simple pay-as-you go access offered at convenient hours has strong popular demand. Without a concerted marketing effort, fee-based access to computers and the Internet are likely to serve only a small segment of the population.

Lessons for those supporting access centres

1. Civil society organizations need to act as watchdogs and place consumer issues related to telecommunications into public debate.

Building telecommunications infrastructure requires huge amounts of capital. Because of this, national telecom services will never operate in an environment of 'true competition', where consumer choice amongst multiple providers promotes efficiency and good practice by competitors. Thus, governments create agencies to regulate the sector. The telecom operators, though, are large and powerful, and the regulators often do not have sufficient political might to manage them. Moreover, the government itself may have stakes in the major operator, as well as the power to override the regulator's decisions. This happened in South Africa on a number of occasions. Such behaviour can result in decisions that do not maximize public interest.

There is clearly a need for third parties which can serve as watchdogs: groups that are not directly engaged in telecom services, not profiting from them, and not directly tied to the government. While the regulator is supposed to play this role, this is demonstrably often not the case, and an outside agitator may help the regulator perform more effectively. In South Africa, for example, the LINK Centre at the University of the Witswatersrand has conducted public-domain research and provided expert commentary on telecommunication regulatory and policy issues. It has done this, for example, for the news media which has in some cases strongly criticized practices of government, the regulator and major operators.

There is an ongoing critical need for third parties to take on this role because of the large vested interests and amounts of money at stake. Major policy decisions are often made behind closed doors and the general public is not usually in much of a position to scrutinize carefully everything that happens. Yet the impacts of these decisions are felt all the way down to access-related activities at the grassroots, which may find themselves struggling simply because they cannot get basic phone service at an affordable price.

2. **Small businesses and not-for-profit groups providing various forms of public ICT access often need support, especially in the form of capital, equipment, practical technical skills, and business management and marketing skills.**

There are plenty of small enterprises struggling to make a success of providing ICT access and related services. Although the capacity of these enterprises in terms of capital, equipment, skills and knowledge varies considerably, most are weak in at least one of the major areas. The idea of applying SMME support services to access centres has a fairly successful history in Asia, mainly through Grameen. In Africa, IDRC supported entrepreneurial start-ups in Senegal and many of these thrived. In the areas covered by this study, however, there was relatively little support activity although some NGOs were beginning to think about it. In the summer of 2003, a project to replicate the success of Bangladesh's Grameen Village Phone Programme was just beginning in northern Uganda, but was too young to be captured in this research.[1] The Vodacom phone shop franchise was another example of an activity fitting this category, with great success. Other licensed national operators were considering how they could emulate the model to meet their own licensing obligations. Clearly, more could be done in this area, especially given the high demand in both countries for productive, financially viable activities. Third parties should help entrepreneurs assess both potential risks and benefits. Market research especially was an area that was often weak, resulting in many stillborn enterprises.

3. **There have been many stand-alone pilot projects and innovative activities, but the knowledge gained from these is fragmented and poorly shared.**

There are many examples of stand-alone initiatives – sometimes implemented by local agencies, sometimes by outsiders – that have shown some level of innovation but which were not covered by this research. Many of them address common problems. For example, Wizzy Digital Courier has developed an innovative solution to the high cost of Internet connections. While they have focused on schools in South Africa, the same idea could be applied to many other situations and places. Sharing these ideas and learning together have proven to be a challenge. Perhaps this is because there has been so much activity around ICTs for development in a reasonably short time period. Many of those who have taken independent action have also expressed some frustration with larger agencies and networks that they have found to be too bureaucratic and often ineffective. This has contributed to overall fragmentation.

For third parties trying to find their own role, there is a rich bank of on-the-ground experience in providing access to ICTs, often for a specific development-related goal. A number of agents have worked at documenting and publicizing these. For example, bridges.org is producing a number of ICT-for-development case studies. At this stage within both South Africa and Uganda, there is little reason for unilateral action. Interventions should be preceded by a concerted effort to understand existing efforts and then designed to complement those activities. The challenge is that the ICT scene is constantly changing and documentation becomes quickly outdated.

4. **Partnerships between national or regional not-for-profit organizations and local access centres require careful attention to the capacity requirements of partners, and may take time to develop as this capacity grows.**

Partnerships between access centres and other organizations were typically ad hoc, sometimes transient, and more common among not-for-profit centres. Such centres, however, often had difficulty maintaining partnerships due to ongoing crises. Some regional structures put in place to support access centres found that maintaining reliable communication with these centres could be difficult; whereas the centres themselves sometimes found such organizations hard to contact and unresponsive to their needs. For example, schools in the Western Cape were unimpressed with the activities of Schoolnet South Africa. So they began their own local network which they hoped to expand. Networking seemed to be more effective where it was rooted in local activities and partnerships expanded slowly, and where the participants had at least some basic potential. External agencies looking to co-ordinate such networks were sometimes far-reaching in their ambitions but had relatively few active linkages to on-the-ground projects.

What are the main lessons for third-party organizations seeking local partners? Firstly, in creating partnerships and networking opportunities, it is best to start small, with concrete activities. These may need to include capacity-building exercises. But if capacity building is not the intention – for example, an access centre is to be used to disseminate information on HIV/AIDS – then the capacity of the centre to effectively carry out such a task should be assessed.

Secondly, not-for-profit organizations are often considered to be prime candidates for partnership, but depending upon the task, small local businesses already providing ICT access or related services may also be valid partners. Partnering with locally owned business has the potential benefit of supporting the local economy while building on existing capacity. Sometimes small businesses exhibit more local ownership than NGOs which may have been opportunistically driven by external funding opportunities. Such businesses also face great challenges in accessing investment capital and sufficient management, market research and technical capacity, and may be equally valid targets for capacity-building efforts.

Research directions

The role of research

The issue of universal access is complex and subject to constant change as new technologies become available and the market changes. For this reason, ongoing research in this area is important. One role of research is to monitor what is happening across countries and to analyse emerging risks and opportunities. This can inform the process of policy revision.

As universal access is a public interest issue, there is a need for independent research and publicly available results. The telecom sector especially has a large body of private sector research often funded by and directed on behalf of major operators. Public research is necessary to inform regulators, smaller operators, consumers and consumer advocates, and to promote transparency. Independent research can also play an important role in monitoring the degree to which reality mirrors policy intent. For instance, in South Africa where there had been a clear discrepancy between policy intent and outcomes, independent research by organizations such as the LINK Centre at the University of Witswatersrand and bridges.org, as well as media coverage of the issue, put pressure on the government to revise its practices.

Much research has been done in this area at the project level and by independent academics, including Master's and doctoral students. While there is a wealth of detail in various case studies, little of this information has been analysed at a more general level. There is potentially much to be learned through meta-analysis. Better channels for sharing and disseminating research on universal access issues would also be useful. Some of the traditional academic methods of disseminating knowledge do not fit well with ICT issues because the information dates very quickly. For this reason, much master's and doctoral research remains unpublished.

The overall arena: actors and relationships

At the national level, ongoing independent research should monitor the impact of universal access policy, with a specific focus on the costs and benefits to various stakeholders. This type of monitoring requires clear definitions of key terms such as 'affordability' so that progress (relative to the definition) can be measured. Other potential areas of research are the effects of pricing structures on consumer behaviour; the impact of specific universal access policies, including the costs and benefits to various stakeholders; international examples and the potential applicability of other policies (for example, those from Latin America) to particular national contexts; and research into the issue of public versus private infrastructure development.

Start-up and sustainability

A key area that remains fuzzy and which is often overlooked is how to determine appropriate strategies for different types of areas. While different strategies are obviously necessary, implementing agencies are usually on their own when it comes to identifying which type of situation they are dealing with. It may be that the market can be left to itself to provide access centres. Or perhaps conditions are such that a subsidy or intervention in setting up a centre would help prime the market. Or it may be a situation where the cost of setting up an access centre outweighs the potential benefits until other conditions – for example, electricity, roads or the local economy – are improved. Research that provides general guidelines and issues to watch out for could help implementing agencies to avoid common errors.

Another area that could be better researched and documented is how best to support entrepreneurs who provide universal access. Some initiatives have already been well documented, including Vodacom's experiences and the Grameen phone programme in Asia. Some needs are already well understood, such as the development of skills in business management, marketing and basic technical maintenance. These particular skills are in high demand. As the importance of entrepreneurs in providing access becomes ever more apparent, these initial case studies and lessons need to be expanded and applied to potential policy in support of entrepreneurial activity. Research can also address the issue of how entrepreneurs make decisions about setting up a business and what types of services to offer. Understanding this could help those who support entrepreneurs to do so in a way that serves the public interest, reduces the risk to entrepreneurs, and maximizes productive potential. For example, a number of entrepreneurs in Lira put great emphasis on customer service and marketing. How could their counterparts in other places be encouraged to adopt such practices and, if they did, would the change be beneficial?

Reach and use

Many not-for-profit telecentres, especially those in or near towns with private competitors, struggle to establish an identity and role for themselves. To survive financially, they often end up competing directly with local businesses on services and prices and targeting the same customers. Further research could identify and document experiences that demonstrate the particular niches such telecentres might fill, ways to extend their reach and development impact, and strategies for survival. And perhaps it could identify when such centres are no longer necessary, in the face of a changing market.

Research on how different people decide on the cost they are willing to pay for a particular service has been rather limited. Largely anecdotal evidence in Africa suggests that ITU guidelines of affordability at 3 per cent of income spent on information and communication services have little connection with reality. In practice, people often pay a much larger proportion of their income on such services, particularly phone use. A related theme is the issue of how people, at the individual and household level, make decisions to start using ICTs and to continue to use them. A special focus on people in the 'subsistence plus' category could lead to greater insight as to when and how to build demand in rural areas. This approach could also be used to analyse differences as a function of languages spoken, sex, age and socio-economic status.

Finally, one of the most important and potentially positive uses of ICTs is to maintain urban-rural and domestic-Diaspora links. Specifically, how much of an equalizing force do these links exert, given other factors like unequal access to education? What are the likely longer-term trends? Ideally, research could inform policies that favour these interactions and thus increase benefits.

Appropriating technologies

Most telecommunications R&D occurs in the industrialized North, targeted towards populations of the same. There is an ongoing need to focus research attention on determining which technologies might be usefully applied to universal access and national development in countries of the South. (GSM, VSAT and wireless are good examples of technologies that have already been of great value to the South.) Ideally, such research should be carried out in a way that supports and strengthens domestic technological capacity, especially so that domestic businesses can participate and gain.

Last words

Optimizing the benefits that new information and communication technologies present to all citizens of a nation is a goal that presents many challenges, constant change and much to be learned. Access centres are an important part of universal access strategies but are not sufficient by themselves. Other conditions are equally crucial, especially effective national telecommunication operators regulated in the public interest. The services that access centres can offer are not always the equivalent of private ownership of ICTs. Sometimes they cannot equal it; sometimes they accomplish more. For example, private cellular phones allow people to receive calls anywhere (within coverage) at any time. But the collaborative dimension of access centres can sometimes help to spark larger community development processes.

One side of universal access that cannot be overlooked is the potential for giving local people opportunities to provide ICT services and profit from that entrepreneurship. When this factor is included, universal access becomes a double-edged tool for community development. Otherwise, if the profit generated from ICT services is sucked out of poor communities and these services do not lead to local economic benefits, universal access is potentially a tool for further impoverishment. Universal access policy aims to maximize benefits and understand and minimize risks to public interest. Early policy, for example in South Africa, was sometimes overly optimistic in identifying potential benefits. Experience now reminds us that positive development outcomes do not spontaneously emerge from access, although the potential is there.

Notes

1 Introduction

1 Not all of the centres included in this study provide full public access, but they serve at least a portion of the public. Schools fit within this understanding, but are included insofar as they provide access to non-students, to maintain manageability of the research. .

2 This presumes the government is democratic and fairly represents the interests of its people.

2 Access centres and South Africa's universal access policy

1 This topic has been well monitored and analysed by various observers, especially the LINK Centre at Wits University. As this overview is comparatively simplistic, the reader is referred to the bibliography for a more detailed account.

2 Because these lines were unaffordable to many households, 2 million were soon disconnected.

3 Uganda's experience with shared access centres and universal access policy

1 It is a 'high risk' proposition mainly because of dependency on third parties for technical solutions and support, difficulty in securing skilled staff, and environmental and climatic conditions. Also, the complexity of factors makes accurate planning difficult. Most external funders require such planning, but provide only short-term funding.

2 In Bukinda, RANET had not been operational long enough for farmers to test this, but in Tororo they reported they had enjoyed benefits due to their ability to respond to the RANET climatic forecasts over three seasons.

3 Bushnet's service has a much broader reach, along the MTN cellular towers that cover the country, although for small cybercafés it is relatively expensive.

4 Social status also carries economic value (although hard to quantify) in terms of favours and privileges others may grant you in the expectation that you will eventually reciprocate.

4   Start-up and scalability of access centres

1   This had been the 'white school' under apartheid, and was still wealthier and preferred by those with a choice, although mixed-race schooling had been introduced.

5   Local livelihoods, reach and development impact

1   The Foodnet project was still fairly new and use was growing – the project also tracks information on use rates. They also printed the prices in newspapers and announced them on the radio.
2   In South Africa, bottle stores sell alcoholic beverages.
3   The Youth Development Trust of South Africa released a report on the topic of computer training, especially at private computer centres, and its links to the job market (YDT, 2003).
4   This is not to suggest that the relatively high cost of telecommunications service in Africa is at all a positive thing! It simply provides evidence by exception that user fees per se are not the sole or even primary reason why computer-based ICTs have much lower usage rates.

6   Conclusions, lessons and research directions

1   Entitled Village Phone, this project is an ongoing partnership between MTN Uganda and the Grameen Technology Center.

Appendix 1: Data Sources and Data Collection Schedule
South Africa

Field research in South Africa took place over a six-and-a-half week period from Tuesday April 22 until Thursday June 5, 2003 and included national level data collection, two community case studies and visits to a number of additional projects. About a third of this time was consumed by travel, logistical preparations, ongoing adjustment of the research design, data analysis and processing and an impressive number of civic holidays. Data collection activities were concentrated in Johannesburg and surrounds, Cape Town, Khayelitsha (on the Cape Flats) and Bhamshela, which is about an hour and a half north-east of Durban.

Research Component	Sub-component	Data collection methods
Khayelitsha Community Case Study	11 ICT access centres, 1 high school, 1 community radio station, 1 computer training school	Observation, Interview with staff and/or management, Half-day user exit surveys at selected locations
	Vulani township tourism project	Meeting/interview w/ technical coordinator
	General population	Survey of 68 households in 3 different areas of Khayelitsha
Bhamshela Case Study	Telecentre	Observation Interviews w/ manager and staff person Group interview w/ students Interviews w/ 2 board members
	Other access centres: • Village Bank • Vodacom container • Telkom container	Observation Interviews w/ management and/or staff Interview w/ board members Half-day user exit surveys
	Qalakale High school	Visit Interview w/ principal
	General population	Survey of 68 households in 3 different communities around Bhamshela

Research Component	Sub-component	Data collection methods
Other sites and projects	Aleksan Kopano Community Resource Centre, Alexandra	Observation, interviews w/ director, director of computer school, youth coordinator, head librarian, and community radio station manager
	Computer Clubhouse, Joburg	Observation, interview w/ manager
	Bridge.Café, Randburg	Observation, interview w/ owner
	Wizzy Digital Courier (multiple locations)	Interviews w/ co-owners, Site visits to schools in Eshowe, Ingwavuma
NGOs and Associations	CINSA/Sangonet	Joint research w/ secretariat (e-mail survey) Participation in conference Monitoring of listserve
	NCRF	Meeting w/ ICT Policy Officer, Communications Officer
	Women's Net	Interview w/ several staff
	Media Works	Interview w/ director, community services coordinator Document review
Governmental agencies	GCIS	Interviews w/ staff Document review
	USA	Interviews w/ head of research department, KZN provincial rep Document review site visits to 2 USA telecentres
Other	Vodacom Community Services	
	Nicholas Pejout, Researcher	Meeting

Uganda

Field research in Uganda lasted seven-and-a-half weeks from Friday June 6 until Wednesday July 30. The research benefited greatly from excellent and very enthusiastic research assistants at each case study site. Very little data processing and analysis took place during this time.

Research Component	Sub-component	Data collection methods
Lira Case Study	CPAR Community Learning Centre	Observation, Interview w/ staff Document review, Interviews w/ 6 participants
	Other access centres (including a school-based telecentre, cybercafés, computer training centres and phone services)	Interviews w/ staff Access centre surveys User exit surveys Observation Student questionnaires
	3 local radio stations	Interview w/ staff Site visits Usage data from household survey
	General population	Survey of about 65 households in Lira and surrounding villages
Kabale Case Study	AHI Telecentres	Document review, Interviews w/ project coordinator and w/ staff at Kabale and Rubaya telecentres
	RANET Bukinda Site	Interview w/ manager, Interview w/ a programme participant
	7 ICT Access Centres (including phone shops, computer training centres, a school-based telecentre)	Interviews w/ staff ICT access centre survey User exit surveys Observation Student questionnaires
	2 local radio stations	Interviews w/ staff, Usage data from household survey
	General population	Survey of 74 households in Bukinda, Rubaya, Kabale town and surrounds
Nabweru Case Study	Nabweru Telecentre	Document review Interview w/ assistant manager Meeting w/ former board member Interviews w/ users (all local gov't representatives) Site visit/observation

Research Component	Sub-component	Data collection methods
Nabweru Case Study	9 ICT Access Centres (including 1 cybercafé, plus phone shops, business centres and computer training centres)	Access centre survey/staff interview Site visits/observation User exit surveys
	General population	Survey of 95 households around Nabweru sub-county
Other locations and projects	Tororo RANET site	Site visit, Interview w/ coordinator, Group interview w/ 6 participants, Document review
	Mbale: radio station, cybercafé	Site visit, interviews w/ management
	Nakaseke telecentre	Site visit, Meeting w/ former manager, current librarian, graduate researcher on-site, administrator, Document review
NGOs and associations	I-Net	Meeting w/ Johnson Nkuuhe, an I-Net founder, Attendance at 2 I-Net workshops, Document review
	Uganda Development Services	Interview w/ founder
	CIAT/AHI	Interview w/ AHI project coordinator, Document review
	Schoolnet	Interview w/ coordinator, Document review
	World Links	Interview w/ coordinator, Document review, Site visits
	Uganda Connects	Interview w/ staff, Document review
	CEEWA	Interview w/ staff, Site visit, Document review
	Wougnet	Interview w/ staff, Document review

Research Component	Sub-component	Data collection methods
NGOs and associations	World Vision	Meeting w/ staff, Site visits to Tororo and Bukinda offices
Government Agencies	UNCST	Interview w/ staff, Document review
	NAADS	Interview w/ staff, Document review
	Dept of Meteorology	Interview w/ RANET project coordinator, Site visits (Tororo, Bukinda), Document review
	Dept of Local Government	Interview w/ director, Site visit to training, Meeting w/ training consultant
	UCC	Interview w/ director, staff, Document review
Other	Bushnet (ISP)	Interview w/ co-founder, Interviews w/ clients & other key informants
	UgandaOnline (ISP)	Interview w/ founder
	UNESCO High Commission for Uganda	Interviews w/ Executive Director, Staff
	Natharius Asingwire, Researcher	Interview
	Isaac Kasana (ICT business sector, independent, previously with Afsat)	Interview
	Gordon Bell (Radio for development)	Interview
	Dick Kawooga (Researcher)	Meeting
	I-Net/APC Policy Workshop	Attendance at four sessions
	AfriNEC meeting	Attendance

Bibliography

Adam, L. (February 2004) Ethiopia Telecommunications Sector Performance Review. University of Addis Ababa, Ethiopia. www.researchictafrica.net/modules.php?op= modload&name=News&file=article&sid=375

Benjamin, P. (ed.) (1998) Multi-Purpose Community Centre Research Report, Version 3.0. National Information Technology Forum. Johannesburg, South Africa. www.sn.apc.org/nitf/mpcc

Benjamin, P. (April 2000) Telecentre 2000 Report 1: Literature Review. LINK Centre, P&DM, WITS University, Pretoria, South Africa.

Benjamin, P. (2001a) Telecentres and Universal Capability. A Study of the Telecentre Programme of the Universal Service Agency, South Africa 1996–2000. Unpublished doctoral dissertation. Aalborg University, Denmark.

Benjamin, P. (2001b) Digital Divide research in South Africa. Paper presented at d3 workshop, Ann Arbor, MI, August 2001.

Benjamin, P. (August 2003) Country profile: South Africa. Unpublished brief prepared for Africa Telecentre Helpnet workshop. CIEUM, Maputo, Mozambique.

Bressie, K., M. Kende and H. Williams. (2004) Telecommunications trade liberalization and the WTO. Paper presented to the 15th ITS Biennial Conference, 'Changing peoples, societies and companies: Telecoms in the 21st Century', Berlin, 5–7 September 2004. http://www.harriswiltshire.com/Telecommunications%20Trade-%20Liberalization%20and%20the%20WTO.pdf

Bridges.org. (2002) Digital Divide Assessment of the City of Cape Town, 2002. Overview of ICT status and Real Access in Cape Town. www.bridges.org/capetown/index.html

Bridges.org. (2003) Telecommunication liberalization – what does it mean for the average citizen? www.bridges.org/e-policy/sa/articles/telcom_liberal.html

British Broadcasting Corporation (BBC). (2002) Universal Service Obligation. T305 – DIGITAL COMMUNICATIONS. Transcript of audio track 5. The Open University, United Kingdom.

Burger, D. (ed.). (2004) South Africa Yearbook 2003/4. GCIS, Pretoria, South Africa. http://www.gcis.gov.za/docs/publications/yearbook04.htm

Chambers, R. (January, 1995) *Poverty and Livelihoods: Whose reality counts?* Institute of Development Studies Discussion Paper No. 347. University of Sussex, United Kingdom.

Community Multimedia Services (CMS) Task Team. (November 2003) Report of the CMS Task Team to the CMS Indaba. Unpublished report, submitted to Department of Communications. Johannesburg, South Africa.

Dagrón, A. (2001) *Making Waves: participatory communication for social change*. The Rockefeller Foundation, New York, USA.

Dymond, A. and S. Oestmann. (2002) *ICTs, Poverty Alleviation and Universal Access: Review of Status and Issues*. ATPS Special Paper Series No. 9. African Technology Policy Studies Network (www.atpsnet.org), Nairobi, Kenya.

Emdon, C. (7 December 2003) Telkom putting a brake on the economy.... A visiting international telecoms specialist and Telkom go head-to-head on telecommunications policy and practice. *Sunday Times*, South Africa. www.suntimes.co.za/2003/12/07/business/companies/comp01.asp

Etta, F.; Parvyn-Wamahiu, S. (ed.). (2003) Information and communication technologies for development in Africa. Volume 2: The experience with community telecentres. International Development Research Centre, Ottowa, Canada.

Freedom of Expression Institute. (February 1998) Universal Service in the Spotlight in South Africa. (In response to Universal Service Agency discussion document) fxi.org.za

FOWODE/Oxfam. (2004) The impact of PMA/NAADS on female subsistence farmers. *New Vision*, 27 July 2004, Kampala, Uganda.

Fuchs, R. (1997) *If You Have a Lemon, Make Lemonade: A Guide to the Start-up of the African Multipurpose Community Telecentre Pilot Projects*. International Development Research Centre, Ottawa, Canada.

Gillwald, A. (2003) *National Convergency Policy in a Globalised World: Preparing South Africa for Next Generation Networks, Services and Regulation*. LINK Centre Public Policy Research Paper No. 4. LINK Centre, Johannesburg, South Africa.

Government Communication and Information System (GCIS). (November, 2001) Multi-purpose Community Centre Business Plan. Formeset Printers, Cape Town, South Africa.

Government of Uganda. (2002) Draft National Information and Communication Technology Policy Framework. Ministry of Works, Housing and Communications, Kampala, Uganda.

Government of Uganda, Ministry of Finance, Planning and Economic Development. (2000) Poverty Eradication Action Plan – Executive Summary. Kampala, Uganda. http://www.finance.go.ug/peap.html

Hanson, H.B. and M. Twaddle (eds.) (1998) *Developing Uganda*. Fountain Publishers, Kampala, Uganda.

Ibrahim, Z. (2004) A word from the trenches. Letter published in *Mail & Guardian*, January 30 to February 5, p. 25. Johannesburg, South Africa.

Jamal, V. (1998) Changes in Poverty Patterns in Uganda. In *Developing Uganda*. Hanson, H.B. and M. Twaddle (eds.). Fountain Publishers, Kampala, Uganda.

Lewis, C. (8 November 2003) Universal Service, Universal Access & the South African Experience. Presented at NetTel Africa Workshop, Abuja, Nigeria. www.ncc.gov.ng/NetTelAfricaWorkshop/TR507_universality-abuja-20031107.ppt

May, J. (ed.). (13 May 1998) Poverty and Inequality in South Africa: Report prepared for the Office of the Executive Deputy President and the Inter-Ministerial Committee for Poverty and Inequality. www.polity.org.za/html/govdocs/reports/poverty.html

Melody, W. (1998) Universal Service in an Information Society. Faculty of Technology, Policy & Management, Delft University of Technology, Delft, The Netherlands.

Melody, W. (1999) Telecom Reform: Progress and Prospects. *Telecommunications Policy*, 23 (1999), pp. 7–34.

Melody, W. (2002) Assessing Telkom's 2003 Price Increase Proposal. LINK Centre Policy Research Paper No. 2. Johannesburg: LINK Centre, University of Witswatersrand.

Ministry of Finance, Planning and Economic Development. (December 2002) Second Participatory Poverty Assessment Report: Deepening the Understanding of Poverty. National Report. Government of Uganda, Kampala, Uganda.

Ministry of Works, Housing and Communications, The President's Office, and National Council for Science and Technology. (July 2002) National Information and Communication Technology Policy. Government of Uganda, Kampala, Uganda.

Msimang, M. (November 2003 draft) Universal Service and Universal Access.

National Agricultural Advisory Service (NAADS). (2001) Programme Implementation Manual NAADS, Kampala, Uganda. http://naads.or.ug/manage/publications/33docpimfinsept.pdf

National Community Radio Forum. (2003) SACRIN Project Profile. NCRF, Johannesburg, South Africa.

Odlyzko, A. (February 2004) Content is Not King. *First Monday*, 6(2).

Odrek Rwabwoogo, M. (2002) *Uganda Districts Information Handbook*. Fountain Publishers, Kampala, Uganda.

Ofir, Z. (January 2003a) Strategic Evaluation: Research Influence on Policy. The Case of South Africa. Final report prepared for the IDRC Evaluation Unit.

Ofir, Z. (January 2003b) Strategic Evaluation: Research Influence on Policy. The Case of Uganda. Final report prepared for the IDRC Evaluation Unit.

Republic of South Africa. (1996a) White Paper on Telecommunications. Government Printers.

Republic of South Africa. (1996b) Telecommunications Act. Government Printers.

Republic of South Africa. (2000) ICASA Act. Government Printers.

Republic of South Africa. (2001) Telecommunications Amendment Act. Government Printers.

Sen, A. (1999) *Development as Freedom*. Oxford Press, London, United Kingdom.

Statistics SA www.statssa.gov.za

Tusubira, F., Kaggwa, I. and F. Mukholi. (2003) Uganda Telecommunications Sector Review. Makerere University, Kampala, Uganda. www.researchictafrica.net/images/upload/Uganda%20SPR%2030_04_04%20v3.pdf

Uganda Bureau of Statistics www.ubos.org

Uganda Communications Commission (UCC). (July 2001) Rural Communications Development Policy for Uganda. Government of Uganda, Kampala, Uganda.

Youth Development Trust. (April 2003) The impact of Information and Communication Technology Training on Youth Entrepreneurship and Job Creation. IDRC, Johannesburg, South Africa. http://www.ydt.co.za/content/pdf/YDTIDRFinalReport3.pdf

Index

Italic page numbers refer to figures and tables.